LIFE CYCLE ASSESSMENT

NEW DEVELOPMENTS AND
MULTI-DISCIPLINARY APPLICATIONS

LIFE CYCLE ASSESSMENT

NEW DEVELOPMENTS AND MULTI-DISCIPLINARY APPLICATIONS

HSIEN HUI KHOO

ICES, A*STAR, Singapore

REGINALD B H TAN

A*STAR, Singapore and National University of Singapore

World Scientific

NEW JERSEY · LONDON · SINGAPORE · BEIJING · SHANGHAI · HONG KONG · TAIPEI · CHENNAI · TOKYO

Published by

World Scientific Publishing Co. Pte. Ltd.

5 Toh Tuck Link, Singapore 596224

USA office: 27 Warren Street, Suite 401-402, Hackensack, NJ 07601

UK office: 57 Shelton Street, Covent Garden, London WC2H 9HE

Library of Congress Cataloging-in-Publication Data

Names: Khoo, Hsien Hui, editor. | Tan, Reginald B. H., editor.
Title: Life cycle assessment : new developments and multi-disciplinary applications /
    Hsien Hui Khoo, ICES, A*STAR, Singapore,
    Reginald B.H. Tan, A*STAR, Singapore and National University of Singapore.
Description: New Jersey : World Scientific, [2022] | Includes bibliographical references and index.
Identifiers: LCCN 2021038193 | ISBN 9789811245794 (hardcover) |
    ISBN 9789811245800 (ebook) | ISBN 9789811245817 (ebook other)
Subjects: LCSH: Environmental management. | Environmental protection.
Classification: LCC GE300 .L54 2022 | DDC 333.7--dc23
LC record available at https://lccn.loc.gov/2021038193

British Library Cataloguing-in-Publication Data
A catalogue record for this book is available from the British Library.

For any available supplementary material, please visit
https://www.worldscientific.com/worldscibooks/10.1142/12515#t=suppl

Desk Editors: Balamurugan Rajendran/Amanda Yun

Typeset by Stallion Press
Email: enquiries@stallionpress.com

Preface

The growing demand for natural resources and products worldwide placed by an ever-increasing population will in turn lead to an intensification in industrial activities. As various types of raw materials are extracted and manufactured, they are accompanied by emissions in the air and water, and direct or indirect impacts on the land. This raises a serious concern in the areas of sustainability science. It becomes clear that an environmental assessment tool is required to aid in making decisions and solving problems related to process or manufacturing systems and their relation to the natural environment.

Often, product or process designs and decision-making are motivated by one single objective, using less energy, reducing waste, or increasing profitability. However, it is necessary to adopt a holistic approach, in which the different parts of the system are assessed simultaneously, considering different types of impact in each part of the process.

Life cycle assessment (LCA) has been internationally accepted as a core topic in the field of environmental management in various industries in order to obtain a complete picture of the environmental impacts of products or processes. In contrast to other types of environmental management tools or sustainability assessment methods, LCA methodologies consider a holistic boundary that includes all relevant processes starting from the extraction of natural resources to various manufacturing stages that lead to the final product. Founded on quantitative science, LCA methodologies have been developed to address real-world, complex problems to uncover "hidden" impacts beyond the conventional boundary of a single-stage manufacturing system to identify sustainable strategies for improvements without burden shifting along the regional or global production chains. Over the years, LCA methodologies have broadened to include life cycle costing (LCC) and social LCA (S-LCA) covering all three dimensions of sustainability (i.e. planet, people, and prosperity). In summary, such developments entail a comprehensive concept known as life cycle sustainability assessment (LCSA).

This book is the first of its kind to offer multi-disciplinary perspectives of new LCA developments and applications — varying from data variability and ecosystem services to the evaluation of the net greenhouse gas emissions from Carbon Capture and Utilization (CCU) methods and waste management. Perspectives of green chemistry principles via LCA and combined life cycle atom economy approaches are explored. In addition, industrial symbiosis concepts, LCA as an Entrepreneurial Tool for Business Management and Green Innovations, and blockchain-enabled LCA are also presented. Land use changes are also considered among the most significant impacts on the environment, particularly on climate, ecosystems, and biodiversity in the field of LCA.

I gratefully acknowledge the Institute of Chemical Engineering and Sciences (ICES), A*STAR, for all the support, research supervision, and opportunities given. My sincere appreciation also goes to the local and international research scientists who contributed to this book. Combined expertise of green chemistry, chemical engineering, process modeling, emerging technological advancements, and economics was applied to expand to the applications and methodology developments in the field of LCA. Among the science experts from A*STAR are Valerio Isoni, Praveen Thoniyot, Pancy Ang, Albertus Handoko, Jonathan Low, and others. My appreciation also goes to Lee Cher Kian, Piya Kerdlap, and Alvin Ee from the National University of Singapore.

With upmost gratitude, I also acknowledge the following (overseas) author contributors who dedicated their time and effort to this book: Van Schoubroeck Sophie (Belgium), Benedetto Rugani (Luxembourg), David Teh (Australia), Arnab Dutta (India), Rodrigo Salvador (Brazil), and Gamini Mendis and Justin Richter (US). Additionally, immense credit goes to Assoc. Professors Miguel Brandão (Stockholm) and Michał Biernacki (Poland), and Prof. Ir. Dr. Suzana Yusup (Malaysia) for sharing their expertise and knowledge to add value to this book.

Most of all, I am extremely grateful to Professor Reginald Tan for providing invaluable guidance and expert advice during my Ph.D. journey at NUS, and most importantly, his professional support for this book.

Last but not least, a special thanks to the WS book project administrator for inviting me to translate my research experience into this multidisciplinary LCA book. I greatly appreciate the WS project team members, Amanda Yun and Balamurugan Rajendran, who have devoted their time to this book.

Hsien Hui, Dr. KHOO
Scientist, ICES, A*STAR

About the Editors

Hsien Hui KHOO is a Research Scientist at the Institute of Chemical & Engineering Sciences (ICES), Agency for Science, Technology and Research (A*STAR), Singapore. She graduated with an MEng degree (Master of Engineering) from Nanyang Technological University in 1998, and completed a research internship for environmental and sustainable management studies at CQU, Australia. In 2007, she obtained her Ph.D. from the National University of Singapore (NUS) specializing on LCA research topics. With a passion for research, she has over 15 years of research experience in the areas of sustainability science and environmental management. Her research topics include sustainability evaluation of processes involving green chemistry, CCU (CO_2 capture and utilization), bio-derived products, waste management, and recycling. She obtained the certificate of *Business Sustainability Management* from the Cambridge Institute of Sustainability Leadership (CISL), UK, in year 2017. Dr. Khoo was also awarded the Singapore Commonwealth Fellowship in Innovation in 2017 by The Royal Commonwealth Society and Enterprise Merit Award in 2018 by Singapore's Quality and Standards Board. In the year 2021, she was invited to be part of the board of Reviewers and Editors for the *Journal of Sustainable Consumption and Production*.

Reginald B. H. TAN is Executive Director at the Science and Engineering Research Council at A*STAR, and concurrently Professor at the Department of Chemical and Biomolecular Engineering at the National University of Singapore.

Prof. Tan has been a keen educator and advocate of science-based and evidence-based solutions to environmental issues and is recognized as a highly respected LCA mentor. He has applied the technique of environmental life cycle assessment to case studies in mineral extraction, energy options, carbon dioxide capture and sequestration, material selection, and waste management, resulting in several significant publications in the field. With his expertise in environmental sustainability, Prof. Tan has been an active standards volunteer over the past 25 years. He has served in many leadership roles both internationally (ISO TC207 SC5 on life cycle assessment and ISO Coordinating Committee on Climate Change) and nationally (Chair of Environmental and Resources Standards Committee). He was awarded the Singapore Standards Council Distinguished Partner Award in 2019.

Contents

https://doi.org/10.1142/9789811245800_0001

Chapter 1

Life Cycle Assessment Methodology: Ongoing Developments and Outlook

Hsien H. KHOO[*,‡] **and Reginald B. H. TAN**[†,§]

*Institute of Chemical and Engineering Sciences (ICES), Agency for Science, Technology & Research (A*STAR), Singapore 627833*

†*Department of Chemical and Biomolecular Engineering, National University of Singapore (NUS), Singapore 117585*

‡*khoo_hsien_hui@ices.a-star.edu.sg*

§*chetanbh@nus.edu.sg*

Life cycle assessment (LCA) has gained worldwide recognition for its use in solving environmental sustainability issues. LCA offers a way to understand the quantitative impact of new processes, products, or services. Its systematic and holistic approach offers the prospect of mapping the series of processes and input–output flows within the defined system boundary of a specific production chain and hence ensures that no "environmental shifting" occurs. This chapter introduces the ISO 14040 standards that define the technical as well as the organizational aspects of LCA phases. The following LCA phases are elaborated: goal and scope definition, inventory analysis, life cycle impact assessment, and finally interpretation. Further methodological developments and emerging areas of applications are discussed. Future-oriented approaches that broaden LCA concepts are also presented.

1. Introduction

Life cycle assessment (LCA) is a technical environmental tool used to help identify the potential "environmental hotspots" which could arise from manufacturing systems or the synthesis of new chemicals. By providing a holistic view of investigations, LCA is capable of considering a set of environmental impact indicators due to the use of different resource or power supplies, thus providing important complementary information in a broad context [1]. LCA methodology involves systematically tracking the list of input–output inventories of a production chain — and their associated emissions released to the natural environment — that leads to the final product of interest [2, 3].

Environmentally friendly or green products are increasing in demand as evidenced by the growth in eco-conscious consumerism and industry [4]. In the modern global economy, international value chains — from raw material extraction and pre-processing to production, use, and disposal of goods — can have varied impacts on the natural environment. LCA can be used to help manage businesses more sustainably and aids in informed decision-making [5, 6]. By quantifying a set of ecological indicators, the generated LCA results can be used to help guide or identify opportunities that deserve further research investigation and at the same time outline strategies that prevent or minimize negative impacts on ecosystems, human health, or natural resources [7]. From an overall perspective, LCA aims to track resource use and their associated pollution that can occur along the production chain of products or services and identify strategies for improvement without burden shifting [8].

2. Life Cycle Assessment Framework: A Revisit to ISO 14040 Series

The Society of Environmental Toxicology and Chemistry (SETAC) [9] started the code of practice and guidelines for the methodological components within LCA in 1993. The set of guidance led to the ISO standardization process for LCA. With the inception of ISO 14040 and ISO 14044 [10−12] international standards, LCA has gained worldwide recognition for its use in solving environmental sustainability issues.

As displayed in Figure 1, the ISO 14040 standards define the technical as well as the organizational aspects of LCA phases. The following standards are included under the 14040 series:

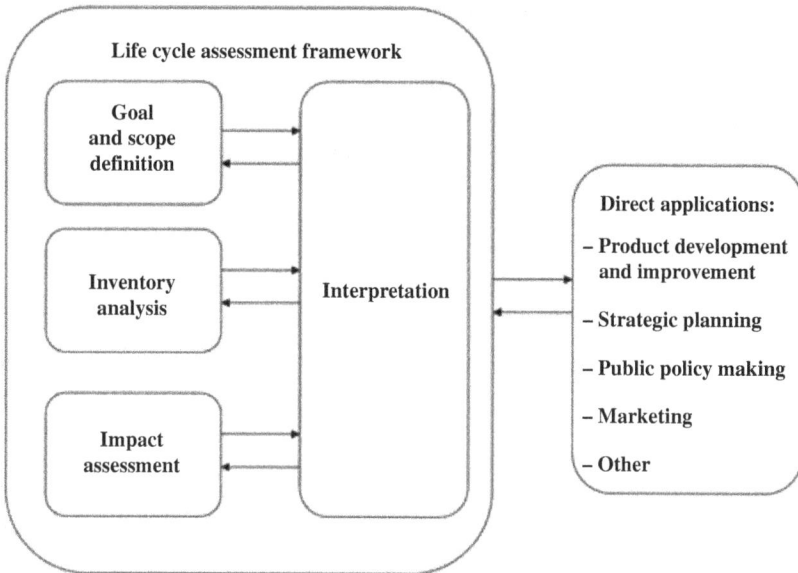

Figure 1.   Phases of a life cycle assessment.

Source: Reproduced from the International Organization for Standardization (ISO). These standards can be obtained from any ISO member and from www.iso.org. Copyright remains with ISO.

- Principles and framework
- Goal and scope definition and inventory analysis
- Life cycle impact assessment
- Interpretation

In order to gain international and stakeholder's acceptance of the LCA framework, Finkbeiner *et al.* [12] reviewed and revised the ISO 14040 and 14044 standards to reaffirm the validity of its main technical contents. Improved readability and the removal of errors and inconsistencies were also achieved for the revised LCA framework.

2.1. *Goal and scope definition phase*

In the beginning stages of an LCA investigation, the product(s) or service(s) to be assessed is defined, a functional unit is chosen, and the required level of details in the LCA model is clearly described. The functional unit of the product or service quantifies the function or service of the system under study and becomes the reference basis to which the inventory analysis and all corresponding LCA results are related. In simplified terms, a common functional unit can be defined as "1 kg of a final product" ready for use. In a more complex large-scale chemical manufacturing scenario, the functional unit may be "100 ton/day of a chemical/solvent". As another example, let's consider an LCA study focusing on the comparison between fossil power plants and renewable energy sources to meet a certain demand, in which the functional unit of the output is "1 MWh" delivered to the electrical grid.

In the goal and scope definition, the boundary or perimeter of the production stages of the series of processes or production chain involved is illustrated. This is typically termed "LCA system boundary" to clarify the list of activities associated with raw material extraction, pre-processing stages, modes of logistics involved, manufacturing of final product, use, and disposal [3, 6]. The scope or system boundaries represent the perimeter of exchanges between a process chain and the environment. Furthermore, assumptions and limitations have to be stated to clearly describe the adopted methodologies for each step of LCA within the system boundary [12, 13]. An LCA "system boundary" has to be defined clearly to illustrate the activities included (or excluded) for the environmental investigation of a product of interest [10, 11]. In common practice, LCA system boundary settings can be defined as follows:

- *Cradle-to-gate*: The series of activities or process stages will start from the extraction of raw materials (e.g. fossil fuels or renewable resources such as biomass) and then proceed to the pre-processing steps before finally reaching the stage of manufacturing of the final product.
- *Gate-to-gate*: This illustrates a limited LCA scope where raw materials/resources for manufacturing are used for production of the final product (the extraction stages of fossil fuels or cultivation activities involved in biomass cultivation are omitted from the system boundary).
- *Cradle-to-grave*: In a complete LCA investigation, the system boundary covers the entire series of activities involved in the extraction of raw materials delivered to the manufacturer for the production of the main product, its use, and final disposal or "end-of-life" stages. The "end-of-life" stages may include the following options: incineration, landfilling, recycling, and remanufacturing/reuse.

In addition, a set of assumptions and operational parameters are also defined [12–15] within each main production stage. Figure 2 illustrates

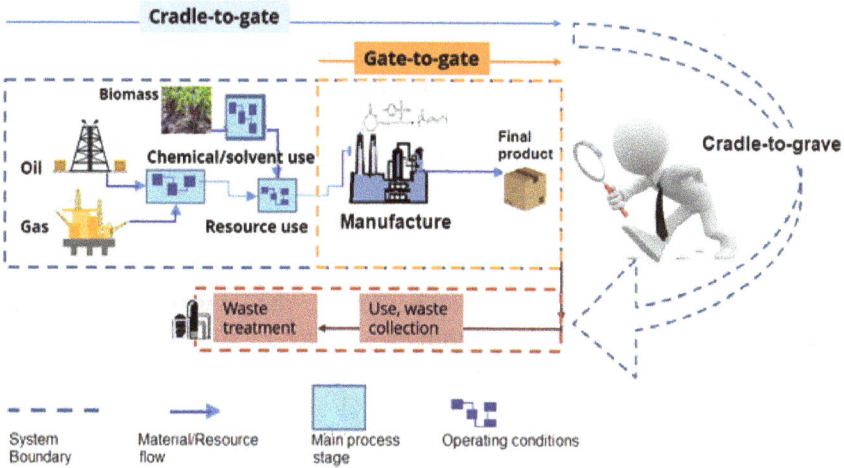

Figure 2. Illustration of LCA system boundary settings.

the system boundary settings for LCA cases of "cradle-to-gate", "gate-to-gate", and "cradle-to-grave".

2.1.1. *System boundary settings*

LCA constitutes a viable screening tool that can pinpoint environmental hotspots in complex value chains, but it is also cautioned that completeness in the scope of investigation has to be appropriately defined [14, 15]. Vigilant system boundary selection for products derived from biomass or agricultural resources is especially important in LCA case studies [16]. The scope of the LCA system boundary should stretch to accommodate farming or agricultural activities since the generation of emissions from water use and fertilizer use, and their associated emissions, occurs at the beginning stages of biomass cultivation.

Limitations and oversimplification of LCA scopes that omit the stages involved in feedstock production can result in unreliable or misleading LCA results [17]. In a noteworthy example, Bernstad Saraiva [18] explored varying choices made in system boundary settings in LCA studies of biorefinery systems. The work demonstrated that the definition of feedstock is of key importance for chosen LCA system boundary settings. Land-use changes (LUC) may also play an imperative aspect in determining the carbon impacts of bio-based fuels [19, 20]. As demonstrated by other examples [21, 22], the internal structure of a process model (e.g. manufacturing operating conditions, technical parameters, and states) within the LCA systems defines the relations between inputs and outputs of the system. In other LCA settings, the comparison of a company that manufactures products overseas versus one that produces the same product locally will have different transport emissions. In such cases, logistical arrangements are included in the LCA system boundary.

2.2. *Inventory analysis phase*

The necessary data and information needed for the inventory and impact assessment are to be identified within the selected LCA system. Life cycle inventory (LCI) is the phase involving quantification of the inputs and outputs for a product throughout its life cycle. Herein, the energy and raw materials used and associated emissions to the atmosphere, water, and soil are quantified for each step in the list of processes defined. Ideally, life cycle inventory accounts for all material and energy flows used in the system. The ISO 14044 series [11] stated ways to manage data quality. The accuracy of LCA results is only as good as the accuracy of the input data. Some of the data requirements that should be addressed are as follows:

- time-relevant and geographical coverage;
- technology information (process operations, scale, yield, etc.);
- representativeness and precision;
- consistency, completeness, and reproducibility.

Additionally, a set of LCI criteria for a multi-disciplinary team (engineers, chemists, technology specialists, etc.) to aid in managing challenges faced in an LCA study carried out for new chemistries or advanced chemical processes was highlighted by Khoo *et al.* [13]. Figure 3 shows the input–output analysis of a "cradle-to-gate" LCA example. Life cycle inventory collection can be a time-consuming and intensive process with strong emphasis on the practice of conducting and organizing a study (data collection, model validation, technical understanding, etc.) [23].

2.2.1. *Data quality requirements*

The accuracy of LCA results is highly dependent on appropriate and reliable data sources. The environmental impact results and conclusions of

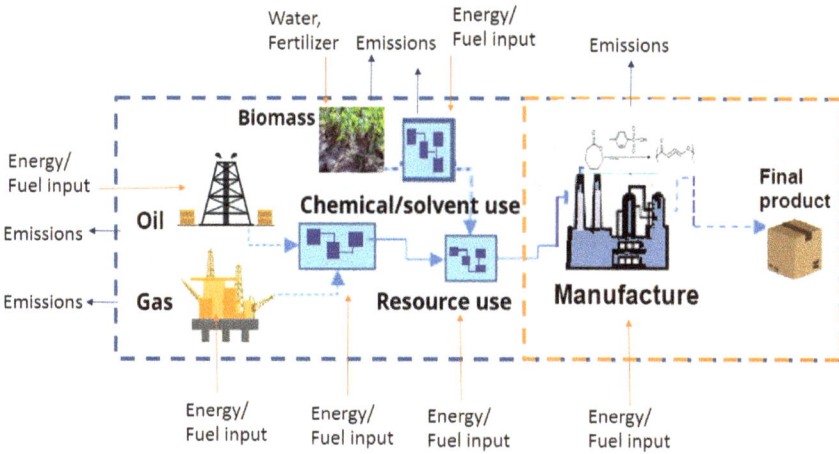

Figure 3.   LCI data involved within an LCA "cradle-to-gate" production system.

LCA investigations are highly dependent on good quality data. Therefore, the sets of data used for constructing LCA models ought to be carefully reviewed. Vigilant practices should be placed on the quality of data used for LCA [24], especially for new synthesis methods which involve complex chemicals or reagents [4, 25]. According to Lewandowska [26], the LCI data collected for LCA studies should be characterized by appropriate quality (spatial, temporal, and technological).

2.3. *Life cycle impact assessment phase*

Life cycle impact assessment (LCIA) evaluates the product life cycle based on the functional unit defined in the LCA system [15, 20]. LCIA refers to the scientific step that converts the LCI data collected to a list of potential environmental impacts [21–23]. In a wide range of LCIA methodologies, various environmental studies have been carried out to generate a variety of characterization factors for different environmental aspects. The impact categories can be related to human health, ecosystems, climate change, and resource depletion. The list of impact categories was selected based on an advanced atmospheric or pollution modeling (fate analysis, exposure, effects, and damage analysis) [26, 27]. A typical structure of the impact assessment method is shown in Figure 4 using IMPACT 2002+ [28] as an example.

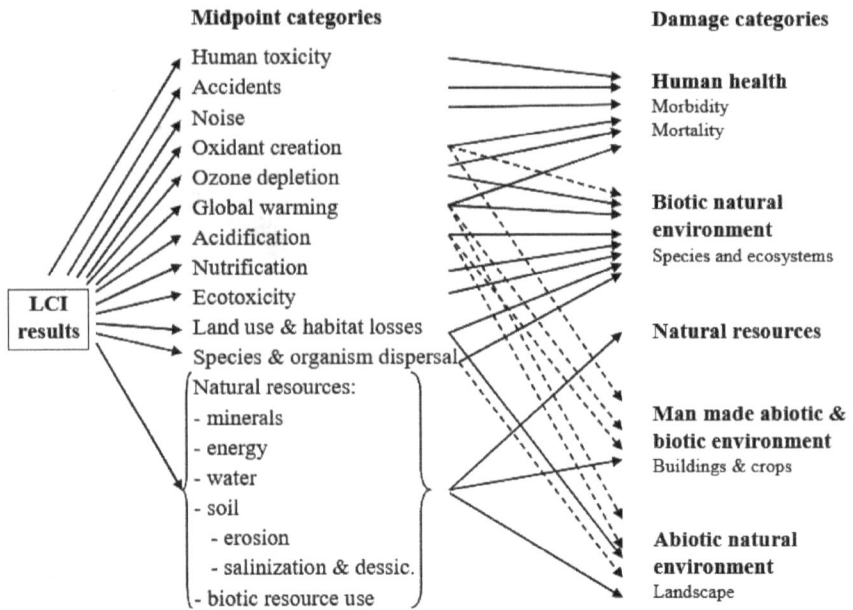

Figure 4.   Overall framework: linking life cycle inventory results via the midpoint impact categories to damage impact categories.

Source: https://lca-net.com/files/LCIA_defStudy_final3c.pdf.

Table 1.   List of LCIA methods.

LCIA Methods	References
IMPACT 2002+	Jolliet *et al.* [28]
CML	Guinée [29]
ReCiPe	Huijbregts *et al.* [30]
EDIP 2003	Hauschild [31]
LIME	Itsubo and Inaba [32]
SIMPASS	Chan *et al.* [33] (in development)

There are numerous impact assessment methods available, each with their own external normalization references. These methodologies provide alternative methods of characterizing and interpreting environmental impacts from the list of chemicals or substances reported in LCI. Table 1 provides some examples of the available LCIA methods.

Various other methodologies have been developed within the domain of LCIA to analyze other types of indicators focusing on resource depletion [34], water footprint or scarcity [35], LUC [36], social impacts in LCA [37], etc.

2.4. *Results interpretation phase*

As described in its framework, applications of LCA focus almost exclusively on eco-efficiency considerations (i.e. resource and emissions intensity per unit good or service) [1–3]. However, measuring environmental impacts is not the only reason why LCA studies are carried out — an important aspect is how the resulting information serves to guide decisions or business strategies. One of the key aims of LCA is to provide the decision makers with comprehensive and scientifically quantified information [13–16]. The outcome of the final LCA step is useful for guiding environmentally friendly decisions [23–26]. This is achieved by proper interpretation of the results generated from an LCA study [38]. A comparative LCA is usually driven by the goal of identifying preferred options (for new technologies or synthesis routes), resulting in the least environmental burden [25, 39]. LCA can be an iterative process; therefore, the interpretation of the LCA can lead to changes in the proposed design, which then leads back to Step 2 in the process (Figure 5).

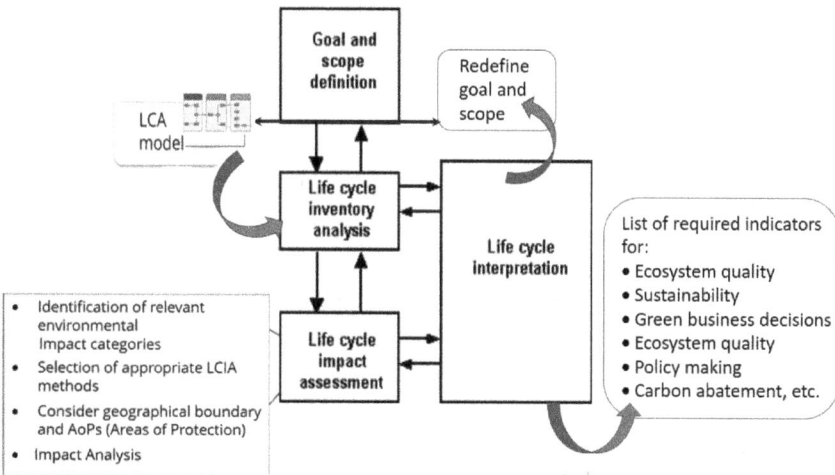

Figure 5.   LCA framework with iterative step.

Recent methodological proposals and developments in the stage of LCA result interpretation aim to broaden the relevance of LCA impact assessment by relating eco-efficiency measures to include planetary or regional sustainability boundaries (e.g. Bjorn and Hauschild [40] and Sandin *et al.* [41]). Pizzol *et al.* [42] seek to improve the interpretation of LCA results derived from LCA studies and communicate such information to stakeholder relevance through the development of normalization and weighting schemes that support prioritizing eco-indicator metrics. Ongoing LCA applications in various emerging sectors are addressed in the following section.

3. Emerging Life Cycle Assessment Applications

The use of LCA has increased rapidly since its conception and is now a well-known environmental assessment tool widely used across various industrial sectors, academia, and other organizations [5, 16, 26]. A wide range of goods and services will have a footprint that can be measured in terms of environmental and ecological consequences. To further seek sustainability measurement indicators, methodological development in LCA has been ongoing with its application expanding in the arena of science and research developments [20, 36, 42]. LCA has found its purpose in analyzing multi-disciplinary areas such as green chemistries, water-energy nexus, ecosystem services, green business models, emerging carbon reduction methods, and so on [43−46].

3.1. *CO_2 reduction technologies*

Rising global concerns around climate change have played a major role in LCA applications. Carbon capture and utilization (CCU) represents a step forward from carbon capture and storage (CCS), as it allows the captured CO_2 to be valorized, repurposed, or reutilized. For an overall sustainability assessment of CCU options, the production chain from CO_2 source to products, including capture technologies, energy and resource requirements, and their associated emissions have to be accounted for. In the pursuit of ensuring proper benefits of CCU, a few applications of LCA in the area of CO_2 valorization have been carried out. Zapp *et al.* [45] demonstrated how LCA is a helpful tool in investigating the environmental consequences associated with carbon reduction technologies. Similarly, LCA was applied by Von der Assen *et al.* [46] to analyze if the use of

CO_2 as a building block for polyurethanes generated environmentally beneficial outcomes or if it will result in higher CO_2 emissions compared with conventional production.

Thonemann and Schulte [47] cautioned that emerging CO_2-based production technologies aim to divulge the possibility of sustainable production in the chemical industry and CCU applications may not lead to improved environmental performance. The authors demonstrated the application of LCA for electrochemical reduction of CO_2 to formic acid from laboratory to industrial scale. Mattia *et al.* [48] echoed the fact that although the production of valuable hydrocarbons from CO_2 could represent the benefit for large emitters to capture the CO_2 for utilization, the potential for mitigating the effects of global warming is not guaranteed. This is due to the use of energy and additional resources required to capture and convert the emitted CO_2 into a useful product. Therefore, a detailed LCA study should be carried out to report realistic net CO_2 results and any other emissions. Figure 6 presents a general description of LCA applied for CO_2 conversion to chemicals and fuels. More of this topic can be found in Chapter 4.

3.2. *Greening of chemical and pharmaceutical industries*

Globally, it is becoming increasingly common for industries to adopt sustainability practices in their manufacturing operations. Both chemical and

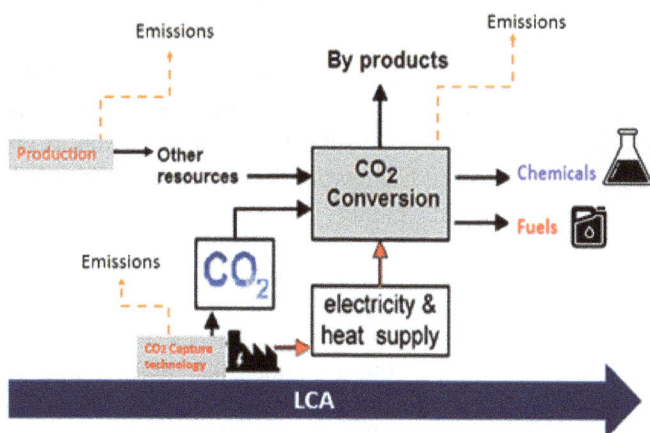

Figure 6.   Application of LCA for CO_2 conversion to chemicals and fuels.

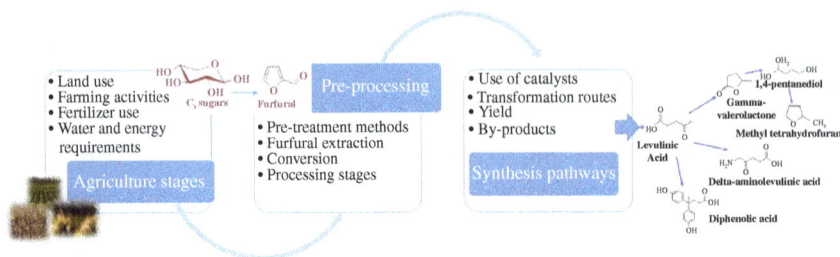

Figure 7.   LCA of biomass to bio-derived chemicals.

pharmaceutical industries are actively taking part in this growing trend. While green chemistry and new bio-derived synthesis pathways have become the forefront of research and development in these industries, LCA has positioned itself as a valuable method to investigate the environmental performance of new chemicals and solvents [4, 16, 25]. LCA studies are able to account for the material flows and fluctuations in the value chain of chemical production pathways. Khoo *et al.* [7] highlighted that bio-based chemicals may not automatically be synonymous with "green" and demonstrated that LCA has been applied to report the environmental performance of bio-products for ensuring their sustainable attributes. The potential LCA steps involved starting from biomass production, pre-processing stages, to the final synthesis pathways of bio-derived chemicals are illustrated in Figure 7.

Anastas and Lankey [49, p. 289] broadened the definition of green chemistry to include *the structure and transformation of all matter* and hazardous impacts of the *full range of threats to human health and the environment*. Kleinekorte *et al.* [50] reiterated that in the search for "greener solutions" in chemistry, the application of LCA is crucial for giving a better understanding of the flow of toxic substances and pollution in production systems by providing a robust framework for gaining new knowledge. Such applications are illustrated in Chapters 5 and 6.

3.3. *Other areas of development*

Suitable methods to address the sustainability of resource use have become an important area of research in sustainability science. LCA is seen to play a principal role in advancing the field of ecosystem services. Klinglmair *et al.* [34] carried out a review of the existing impact assessment methods dealing with resources and identified key areas of improvement for more robust and comprehensive methods for evaluating resources in the LCA context.

In alternative areas, emerging LCA development focuses on freshwater scarcity in many regions of the world. Freshwater use and scarcity are expected to become one of the most sensitive environmental issues in the coming decades, however, most existing LCA methodologies generally do not include impact assessment models for addressing the potential environmental impacts of freshwater use or depletion. The aim to develop quantitative of the cause–effect chain relationship of water use has been presented by Bayart *et al.* [35].

As a final example as well as the revision of the LCA framework [10, 11], it was described in Section 2.2.1 that LCA system boundaries which neglect the use of biomass feedstock production can result in unreliable or misleading LCA results [17]. It is highlighted (again) that this is especially essential for bio-based products or fuels [18]. One of the key topics within the arena of biomass utilization covers the impact of LUC in LCA. The growing requirement for the use of land to meet increasing demands for biodiesel has significant impacts on ecosystem quality and biodiversity. New methodological developments within the environmental LCIA models aimed at establishing a cause–effect relationship between the demand for biodiesel and its impacts on land use are ongoing (e.g. [17, 20, 36]). With evolving bio-based plastics and fuels, the potential environmental performance of new renewable products has become a crucial aspect for different stakeholders. Consequential and attributional approaches to LCA for such subjects will play vital roles in system boundary settings [51, 52].

4. Outlook: Future-Oriented Life Cycle Assessment Concepts

As LCA evolves to evaluate new challenges faced in increased complexities of processes or global production supply chains, LCA applications are expected to keep pace with changes in new sustainability objectives. The broadening of LCA methodologies may include risk assessment combined with LCA. In another area of advancing LCA concepts, the parameters of dynamic, temporal aspects of manufacturing systems, or supply chains are expected to be widely addressed. New technological advancements such as blockchain concepts are also making their way into the future of LCA.

4.1. *Risk assessment and life cycle assessment*

Matthews *et al.* [53] noted that the current LCA practices are not able to provide risk analysts with detailed information related to chemical characteristics and any potential discharges that may occur and have negative effects on individuals due to their toxic effects. It was suggested that in order to provide valuable information to decision makers, advances in both LCA and Risk Assessment (RA) methods are required. However, LCA seeks a holistic approach for the generation of a list of impacts across multiple environmental categories, and on the other hand, RA focuses on the management of a toxicological hazard in specific exposure scenarios. The integration of RA–LCA is not an easy task as different disciplines of knowledge are required. Challenges in choosing the appropriate data measures and selection matters arise while using the RA–LCA method [54, 55].

Walser *et al.* [56] proposed a method for the combination of LCA, RA, and human biomonitoring to improve regulatory decisions and policy making for chemicals. The authors claim that the combined use of these three methods allows a robust search for sustainable alternatives of the currently marketed chemicals that have unfavorable risk profiles. In another example, Pennington *et al.* [57] suggested comparing quantities of chemicals released into the environment in terms of the risk and consequences of toxicological effects in LCA investigations. The authors provided a step-by-step description of the methodological similarities and differences between risk- and hazard-based indicators for LCA applications.

4.2. *Dynamic life cycle assessment*

In most conventional LCA models, the entire production chain is considered as a "static", non-dynamic system, with its corresponding input–output data occurring at one point in time; therefore, the generated impact results do not factor in any time-related changes in the environment [5, 15, 24]. In order to expand the horizons of LCA, Beloin-Saint-Pierre *et al.* [58] proposed an approach to incorporate time dimensions in LCI (input–output) data. Known as the Enhanced Structural Path Analysis, the method considers environmental interventions which are distributed over a time frame. In another example, Tiruta-Barna *et al.* [59] presented a dynamic method which deals with complex supply chains and related

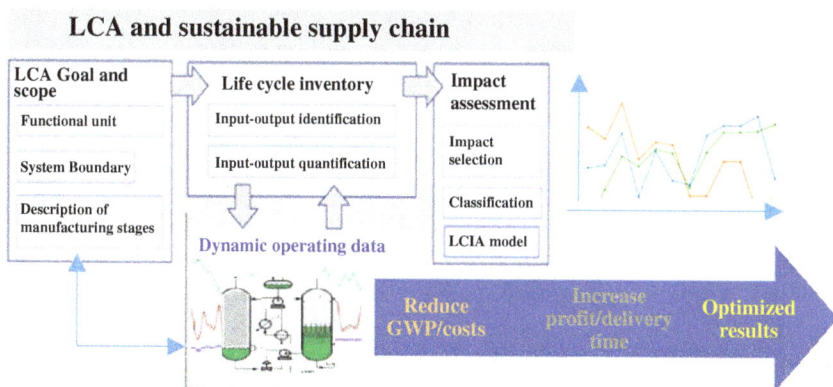

Figure 8.   Proposed dynamic LCA–SC model.

Source: Adapted from Khoo *et al.* [8].

processes presented in LCA studies. Yet another dynamic LCA approach was proposed by Levasseur *et al.* [60] to consider the temporal profile of emissions in LCI. In their analysis, time-dependent characterization factors were calculated to assess real-time impact scores for any given time horizon.

In the domain of evolving dynamic LCA concepts, a noteworthy example was given by Pigné *et al.* [61]. Embedded in the LCA model are unit process operations fluctuating over time, with exchange of mass and energy supplies depending on production and supplies described by temporal parameters and functions. In a combined LCA–Supply Chain (SC) modeling work for biomass production to bio-based chemicals, Khoo *et al.* [8] highlighted that several multi-scaled factors concerning the sustainability outcome of LCA–SC modeling approaches still deserve further investigation efforts. This is due to complex SC networks with multiple measures that characterize their performances. A proposed dynamic LCA–SC model is illustrated in Figure 8.

4.3. *Blockchain-enabled life cycle assessment*

One of the major challenges in LCA is collecting and tracking reliable input–output data [13]. This time-consuming effort can be compounded by complexities involved in quantifying inputs and outputs at multiple supply chain stages [8, 20, 24]. The challenge of assessing emerging technologies to aid data tracking has been increasingly discussed in the field of LCA. Future-oriented LCA studies will be expected to be coupled with technological innovation. Zhang *et al.* [62] suggested that blockchain technology should be offered as an "ideal solution" to overcome the challenges associated with obtaining and tracking appropriate input–output data in value chains. The authors claimed that with the adoption of blockchain technology, all the data on input substances and output emissions can be readily shared and integrated to evaluate the environmental performance of a product life cycle. The adoption of blockchain technology can also enhance green and sustainable practices in supply chains [63, 64]. In a final example, Teh *et al.* [65] showed that blockchain-enabled LCA systems can be used to support sustainability and help organizations achieve targeted Sustainable Development Goals (SDGs). This is shown in Chapter 10.

5. Concluding Remarks

The awareness that all types of goods and services will have a footprint that can be measured in terms of environmental/ecological consequences has gained attention. Henceforth, LCA is an important concept to be employed for companies that seek to enter the era of green business markets [1–5]. LCA offers a way to understand the quantitative impact of new processes, products, or services. If real environmental improvements are to be made by changes in the product or service, it is important not to cause greater environmental deteriorations at another time or place in the product's life cycle. The key benefit of conducting LCA is to gain the full picture of the impacts of a product, process, or activity and find the best solutions for improvement. While the overall goal of the assessment is to review environmental impacts, social and economic impacts can also be considered in the investigation [8, 26, 31].

As a powerful tool which helps manage businesses more sustainably, LCA has gained international recognition for its use in informed decision-making for the production of "greener" chemicals or products [16–18]. Its systematic and holistic approach offers the prospect of mapping the series of processes and input–output flows within the defined system boundary (e.g. energy and resource use, solid wastes, and emissions) of an entire production chain and hence ensures that no "environmental shifting" occurs [13–16]. By quantifying a set of ecological indicators within a defined LCA model, the generated LCA results can be used to help guide or identify opportunities that deserve further research investigation [21, 31, 36]. LCA has also gained its importance in driving the development of products and processes with improved environmental credentials at an early research stage [45−48]. Future developments that expand the horizons of LCA methodologies can be found in the areas of RA [55, 56], dynamic operations or supply chain performances [58−60], and finally blockchain technology [63−65].

References

[1] J. R. Stewart, M. W. Collins, R. Anderson, and W. R. Murphy, Life Cycle Assessment as a tool for environmental management. *Clean Prod. Process.* **1**, 73–81 (1999).

[2] M. Finkbeiner, M. Wiedemann, and K. Saur, A comprehensive approach towards product and organization related environmental management tools. *Int. J. Life Cycle Assess.* **3**, 169 (1998).

[3] A. Tukker, Life Cycle Assessment as a tool in environmental impact assessment. *Environ. Impact Assess. Rev.* **20**, 435–456 (2000).

[4] M. Secchi, V. Castellani, E. Collina, N. Mirabella, and S. Sala, Assessing eco-innovations in green chemistry: Life Cycle Assessment (LCA) of a cosmetic product with a bio-based ingredient. *J. Clean. Prod.* **129**, 269–281 (2016).

[5] J. B. Guinée, R. Heijungs, G. Huppes, A. Zamagni *et al.*, Life Cycle Assessment: Past, present, and future. *Environ. Sci. Technol.* **45**(1), 90–96 (2011).

[6] G. Finnveden, M. Z. Hauschild, T. Ekvall, J. Guinée, R. Heijungs *et al.*, Recent developments in life cycle assessment. *J. Environ. Manage.* **91**(1), 1–21 (2009).

[7] H. H. Khoo, L. L. Wong, and J. Tan *et al.*, Synthesis of 2-Methyl tetrahydrofuran from various lignocellulosic feedstocks: Sustainability assessment via LCA. *Res. Conserv. Rec.* **95**, 174–182 (2015).

[8] H. H. Khoo, R. M. Eufrasio-Espinosa, and L. S. C. Koh *et al.*, Sustainability assessment of biorefinery production chains: A combined LCA-supply chain approach. *J. Clean. Prod.* **235**, 1116–1137 (2019).

[9] W. Klöpffer, The role of SETAC in the development of LCA. *Int. J. Life Cycle Assess.* **11**, 116–122 (2006).

[10] ISO 14040, *Environmental Management: Life Cycle Assessment — Principles and Framework* (International Organization for Standardization, Geneva, 2006).

[11] ISO 14044, *Environmental Management: Life Cycle Assessment — Requirements and Guidelines* (International Organization for Standardization, Geneva, 2006).

[12] M. Finkbeiner, A. Inaba, R. B. H. Tan, K. Christiansen, and H.-J. Klüppel, The new international standards for life cycle assessment: ISO 14040 and ISO 14044. *Int. J. Life Cycle Assess.* **11**, 80–85 (2006).

[13] H. H. Khoo, V. Isoni, and P. N. Sharratt, LCI data selection criteria for a multidisciplinary research team: LCA applied to solvents and chemicals. *Sustain. Prod. Consum.* **16**, 68–87 (2018).

[14] S. Suh, M. Lenzen, G. J. Treloar, H. Hondo, A. Horvath, G. Huppes *et al.*, System boundary selection in life-cycle inventories using hybrid approaches. *Environ. Sci. Technol.* **38**(3), 657–664 (2004).

[15] T. Li, H. Zhang, Z. Liu, Q. Ke, and L. Alting, A system boundary identification method for life cycle assessment. *Int. J. Life Cycle Assess.* **19**, 646–660 (2014).

[16] H. H. Khoo, W. L. Ee, and V. Isoni, Bio-chemicals from lignocellulose feedstock: Sustainability, LCA and the green conundrum. *Green Chem.* **18**, 1912–1922 (2016).

[17] T. Searchinger, R. Heimlich, R. A. Houghton, F. Dong, A. Elobeid, J. Fabiosa *et al.*, Use of U.S. croplands for biofuels increases greenhouse gases through emissions from land-use change. *Science* **319**, 1238–1240 (2008).

[18] A. Bernstad Saraiva, System boundary setting in life cycle assessment of biorefineries: A review. *Intl. J. Environ. Sci. Technol.* **14**, 435–452 (2017).

[19] J. Fargione, J. Hill, D. Tilman, S. Polasky, and P. Hawthorne, Land clearing and the biofuel carbon debt. *Science* **319**, 1235–1238 (2008).

[20] D. Tonini, L. Hamelin, and T. Astrup, Environmental implications of the use of agro-industrial residues for biorefineries: Application of a deterministic model for indirect land-use changes. *GCB Bioenergy* **8**, 690–706 (2016).

[21] B. Löfgren and A.-M. Tillman, Relating manufacturing system configuration to life-cycle environmental performance: Discrete-event simulation supplemented with LCA. *J. Clean. Prod.* **19**(17–18), 2015–2024 (2011).

[22] K. Harun and K. Cheng, Life Cycle Simulation (LCS) approach to the manufacturing process design for sustainable manufacturing. *2011 IEEE International Symposium on Assembly and Manufacturing (ISAM)*, 1–8 (2011).

[23] S. Neugebauer, J. Martinez-Blanco, R. Scheumann, and M. Finkbeiner, Enhancing the practical implementation of life cycle sustainability assessment — proposal of a Tiered approach. *J. Clean. Prod.* **102**, 165–176 (2015).

[24] D. Cespi, E. S. Beach, T. E. Swarr, F. Passarini, I. Vassura *et al.*, Life cycle inventory improvement in the pharmaceutical sector: Assessment of the sustainability combining PMI and LCA tools. *Green Chem.* **17**, 3390–3400 (2015).

[25] D. Ott, S. Borukhova, and V. Hessel, Life cycle assessment of multi-step rufinamide synthesis — from isolated reactions in batch to continuous microreactor networks. *Green Chem.* **18**, 1096–1116 (2016).

[26] A. Lewandowska, Environmental life cycle assessment as a tool for identification and assessment of environmental aspects in environmental management systems (EMS) part 1: Methodology. *Int. J. Life Cycle Assess.* **16**, 178–186 (2011).

[27] M. Z. Hauschild, M. Goedkoop, J. Guinée, R. Heijungs, M. Huijbregts, O. Jolliet *et al.*, Identifying best existing practice for characterization modeling in life cycle impact assessment. *Int. J. Life Cycle Assess.* **18**(3), 683–697 (2013).

[28] O. Jolliet, A. Brent, M. Goedkoop, N. Itsubo, R. Mueller-Wenk, C. Peña *et al.*, *LCIA Definition Study of the SETAC-UNEP Life Cycle Initiative* (United Nations Environment Programme, 2003).

[29] J. B. Guinée, *Handbook on Life Cycle Assessment: Operational Guide to the ISO Standards* (Kluwer Academic Publishers, Dordrecht, Boston, 2002).

[30] M. A. J. Huijbregts, Z. J. N. Steinmann, P. M. F. Elshout, G. Stam, F. Verones, M. Vieira *et al.*, ReCiPe2016: A harmonised life cycle impact assessment method at midpoint and endpoint level. *Int. J. Life Cycle Assess.* **22**, 138–147 (2017).

[31] M. Hauschild, Spatial differentiation in life cycle impact assessment: A decade of method development to increase the environmental realism of LCIA. *Int. J. Life Cycle Assess.* **11**, 11–13 (2006).

[32] N. Itsubo and A. Inaba, A new LCIA method: LIME has been completed. *Intl. J. Life Cycle Assess.* **8**, 305–305 (2003).

[33] Y. T. Chan, R. B. H. Tan, and H. H. Khoo, Characterisation framework development for the SIMPASS methodology. *Int. J. Life Cycle Assess.* **17**(1), 89–95 (2012).

[34] M. Klinglmair, S. Sala, and M. Brandão, Assessing resource depletion in LCA: A review of methods and methodological issues. *Int. J. Life Cycle Assess.* **19**, 580–592 (2014).

[35] J.-B. Bayart, C. Bulle, L. Deschênes, M. Margni, S. Pfister, F. Vince *et al.*, Framework for assessment of off-stream freshwater use within LCA. *Int. J. Life Cycle Assess.* **15**, 439–453 (2010).

[36] J. H. Schmidt, P. Christensen, and T. S. Christensen, Assessing the land use implications of biodiesel use from an LCA perspective. *J. Land Use Sci.* **4**, 35–52 (2009).

[37] L. C. Dreyer, M. Z. Hauschild, and J. Schierbeck, Characterisation of social impacts in LCA. *Int. J. Life Cycle Assess.* **15**, 247–259 (2010).

[38] V. Prado, B. A. Wender, and T. P. Seager, Interpretation of comparative LCAs: External normalization and a method of mutual differences. *Int. J. Life Cycle Assess.* **22**, 2018–2029 (2017).

[39] J. C. Powell, D. W. Pearce, and A. L. Craighill, Approaches to valuation in LCA impact assessment. *Int. J. Life Cycle Assess.* **2**, 11–15 (1997).

[40] A. Bjorn and M. Hauschild, Introducing carrying capacity-based normalisation in LCA: Framework and development of references at midpoint level. *Int. J. Life Cycle Assess.* **20**(7), 1005–1018 (2015).

[41] G. Sandin, G. Peters, and M. Svanstrom, Using the planetary boundaries framework for setting impact-reduction targets in LCA contexts. *Int. J. Life Cycle Assess.* **20**(12), 1684–1700 (2015).

[42] M. Pizzol, A. Laurent, S. Sala, B. Weidema, F. Verones, and C. Koffler, Normalization and weighting in life cycle assessment: *quo vadis? Int. J. Life Cycle Assess.* **22**, 853–866 (2017).

[43] V. K. K. Upadhyayula, D. E. Meyer, M. A. Curran, and M. A. Gonzalez, Life cycle assessment as a tool to enhance the environmental performance of carbon nanotube products: A review. *J. Clean. Prod.* **26**, 37–47 (2012).

[44] N. Thonemann, A. Schulte, and D. Maga, How to conduct prospective life cycle assessment for emerging technologies? A systematic review and methodological guidance. *Sustainability* **12**(3), 1192 (2020).

[45] P. Zapp, A. Schreiber, J. Marx, M. Haines, J.-F. Hake, and J. Gale, Overall environmental impacts of CCS technologies — A life cycle approach. *Int. J. GHG Contrl.* **8**, 12–21 (2012).

[46] N. von der Assen, P. Voll, M. Peters, and A. Bardow, Life cycle assessment of CO_2 capture and utilization: A tutorial review. *Chem. Soc. Rev.* **43**, 7982–7994 (2014).

[47] N. Thonemann and A. Schulte, From laboratory to industrial scale: A prospective LCA for electrochemical reduction of CO_2 to formic acid. *Environ. Sci. Technol.* **53**, 12320–12329 (2019).

[48] D. Mattia, M. D. Jones, J. P. O'Byrne, O. G Griffiths, R. E. Owen, E. Sackville *et al.*, Towards carbon neutral CO_2 conversion to hydrocarbons. *ChemSusChem* **8**, 4064–4072 (2015).

[49] P. T. Anastas and R. L. Lankey, Life cycle assessment and green chemistry: The Yin and Yang of industrial ecology. *Green Chem.* **2**, 289–295 (2000).

[50] J. Kleinekorte, L. Fleitmann, M. Bachmann, A. Kätelhön, A. Barbosa-Póvoa, N. von der Assen *et al.*, Life cycle assessment for the design of chemical processes, products, and supply chains. *Annu. Rev. Chem. Biomolecular Eng.* **11**(1), 203–233 (2020).

[51] V. Venkatachalam, S. Spierling, R. Horn, and H.-J. Endres, LCA and eco-design: Consequential and attributional approaches for bio-based plastics. *Procedia CIRP* **69**, 579–584 (2018).

[52] A. Roos and S. Ahlgren, Consequential life cycle assessment of bioenergy systems — A literature review. *J. Clean. Prod.* **189**, 358–373 (2018).

[53] H. S. Matthews, L. Lave, and H. MacLean, Life cycle impact assessment: A challenge for risk analysts. *Risk Analysis: Int. J.* **22**, 853–860 (2002).

[54] I. Linkov, B. D. Trump, B. A. Wender, T. P. Seager, A. J. Kennedy, and J. M. Keisler, Integrate life-cycle assessment and risk analysis results, not methods. *Nature Nanotech.* **12**, 740–743 (2017).

[55] S. J. Cowell, R. Fairman, and R. E. Lofstedt, Use of risk assessment and life cycle assessment in decision making: A common policy research agenda. *Risk Anal.* **22**, 879–894 (2002).

[56] T. Walser, R. M. Bourqui, and C. Studer, Combination of life cycle assessment, risk assessment and human biomonitoring to improve regulatory decisions and policy making for chemicals. *Environ. Impact Assess. Rev.* **65**, 156–163 (2017).

[57] D. W. Pennington, M. Margni, J. Payet, and O. Jolliet, Risk and regulatory hazard-based toxicological effect indicators in Life-Cycle Assessment (LCA). *Human Ecol. Risk Assess: Int. J.* **12**, 450–475 (2006).

[58] D. Beloin-Saint-Pierre, R. Heijungs, and I. Blanc, The ESPA (enhanced structural path analysis) method: A solution to an implementation challenge for dynamic life cycle assessment studies. *Int. J. Life Cycle Assess.* **19**, 861–871 (2014).

[59] L. Tiruta-Barna, Y. Pigné, T. N. Gutiérrez, and E. Benetto, Framework and computational tool for the consideration of time dependency in life cycle inventory: Proof of concept. *J. Clean. Prod.* **116**, 198–206 (2016).

[60] A. Levasseur, P. Lasage, M. Margini, L. Deschênes, and R. Samson, Considering time in LCA: Dynamic LCA and its application to global warming impact assessments. *Environ. Sci. Technol.* **44**(8), 3169–3174 (2010).

[61] Y. Pigné, T. N. Gutiérrez, T. Gibon, T. Schaubroeck, E. Popovici, A. Hayato *et al.*, A tool to operationalize dynamic LCA, including time differentiation on the complete background database. *Int. J. Life Cycle Assess.* **25**, 267–279 (2020).

[62] A. Zhang, R. Y. Zhong, M. Farooque, K. Kang, and V. G. Venkatesh, Blockchain-based life cycle assessment: An implementation framework and system architecture. *Res. Conserv. Rec.* **152**, 104512 (2020).

[63] S. Saberi, M. Kouhizadeh, J. Sarkis, and L. Shen, Blockchain technology and its relationships to sustainable supply chain management. *Int. J. Prod. Res.* **57**, 2117–2135 (2019).

[64] M. Kouhizadeh and J. Sarkis, Blockchain practices, potentials, and perspectives in greening supply chains. *Sustainability* **10**(10), 3652 (2018).

[65] D. Teh, T. Khan, B. Corbitt, and C. E. Ong, Sustainability strategy and blockchain-enabled life cycle assessment: A focus on materials industry. *Environ. Sys. Dec.* **40**, 605–622 (2020).

© 2022 World Scientific Publishing Company
https://doi.org/10.1142/9789811245800_0002

Chapter 2

Life Cycle Inventory Data Variability: Case of Tetrahydrofuran Manufacture

Lee CHER KIAN*, Reginald B. H. TAN*,‡ and Hsien H. KHOO†,§

*Department of Chemical & Biomolecular Engineering, National University of Singapore (NUS) Singapore 117585

†Institute of Chemical and Engineering Sciences (ICES), Agency for Science, Technology & Research (A*STAR), Singapore 627833

‡chetanbh@nus.edu.sg

§khoo_hsien_hui@ices.a-star.edu.sg

Life cycle inventory (LCI) is an imperative aspect of life cycle assessment (LCA) where the procedure of data collection involving the energy and materials used as well as the by-products and wastes released — throughout the production chain of a product — is systematically quantified. This step, which is the fundamental building block of information needed to model LCA systems, will require a substantial amount of data. A process-based approach is introduced to help with the generation of data required for the LCA comparisons. Necessary steps for proper management of LCA data requirements were also discussed.

1. Background: Life Cycle Assessment Data Requirements

As an environmental management tool that provides a methodological framework for evaluating the potential environmental impacts attributable to the life cycle of a product system, LCA has positioned itself as a valuable method in the chemical and pharmaceutical industries. Life cycle inventory (LCI) is an imperative aspect of LCA where the procedure of data collection involving energy and materials used, by-products and wastes released — throughout the production chain of a product — is systematically quantified [1]. This step, which includes the fundamental building blocks of information needed in LCA (illustrated in Figure 1), requires large amounts of data that are often challenging to collate.

In this chapter, a process-based approach is introduced to help with the generation of data required for the LCA comparisons. The involvement and knowledge of the process engineers may be required to consider model-based estimation and analysis of energy utilities for industrial manufacturing of chemicals [2]. Herein, data variability is elaborated in the scope of process-based data generation. Material input–output data will be obtained from process-based information extracted from patents and reports. Issues of data variability, which affect the overall LCA results, will be discussed.

2. Case Study: Life Cycle Assessment of Tetrahydrofuran Manufacture

The production of tetrahydrofuran (THF) is selected for investigation. The Reppe process, which uses acetylene and formaldehyde as its starting materials, is the dominant method of the 1,4-Butanediol (BDO)

Figure 1.   Steps involved: From LCI to LCA modeling.

production process and is heavily employed by producers like BASF and INVISTA [3]. Dairen Chemicals produces BDO (1,4-butanediol) using propylene, while a handful of companies use MAH (maleic anhydride) to produce THF, BDO, and gamma-butyrolactone (GBL) through a process commonly known as the Davy process. Mitsubishi Chemicals uses its technology to convert butadiene into BDO and THF [4].

The following four cradle-to-gate industrial production routes for the manufacturing THF are investigated:

(i) Reppe (the Reppe route is not be confused with the Reppe process)
(ii) Butadiene
(iii) Propylene (PP)
(iv) Maleic anhydride (MAH)

The four production routes — differentiated by their starting input materials — are summarized in Table 1.

2.1. *Life cycle assessment system boundary*

The basic concept of LCA is to collect all mass and energy flows that are associated with the production chain of a product and to translate the resulting inventory into environmental impacts. The system boundary defines the production stages that are included in the assessment.

The LCA boundary starts from raw materials extraction and ends at the finished THF product. For each option, the main production stages are sectioned into unit processes to track the transformation of various starting materials into THF, as shown in Figures 2(a)–(d).

2.2. *Life cycle inventory data*

As part of the chapter's objective, the input–output material data collection approach basically consists of three main steps, as illustrated in Figure 3. The steps start with the sourcing of information such as yield, selectivity, plant layout, stream data and operating conditions, from scientific articles, reports, and patents. In order to represent the orders of data reliability, the steps include a flowchart with three feedback loops coded in green, orange, and red boxes in the order of *high → moderate → low*, respectively. The details of data reliability, and their respective color codes, are exemplified in Table 2.

Table 1.   THF industrial production routes.

(i) Reppe Process Route	
Main Process Steps	Chemical Transformation
Ethynylation reaction of formaldehyde	$C_2H_2 + 2CH_2O \rightarrow$ 1,4-butynediol
Hydrogenation of But-2-yne-1,4-diol	1,4-butynediol + $2H_2 \rightarrow$ 1,4-butanediol
Cyclization of BDO	1,4-butanediol $\rightarrow$ THF + H_2O

(ii) Butadiene Process Route	
Main Process Steps	Chemical Transformation
Acetoxylation of Buta-1,3-diene	1,3-butadiene + 2 Acetic acid + $0.5\ O_2 \rightarrow$ 1,4-diacetoxy-2-butene + H_2O
Hydrogenation of 1,4-diacetoxybutene	1,4-diacetoxy-2-butene + $H_2 \rightarrow$ 1,4-diacetoxy-2-butane
Hydrolysis and cyclization of 1,4-diacetoxybutane	1,4-diaceoxybutane + $H_2O \rightarrow$ aceoxyhydroxylbutane + AcOH Aceoxyhydroxylbutane + H2O $\rightarrow$ 1,4-butanediol + AcOH Aceoxyhydroxylbutane $\rightarrow$ THF + AcOH 1,4-butanediol $\rightarrow$ THF + H_2O

(iii) Propylene Process Route	
Main Process Steps	Chemical Transformation
Propylene acetylation	Propylene + ½ O_2 + Acetic Acid $\rightarrow$ Allyl acetate + H_2O
Hydrolysis of allyl acetate	Allyl acetate + $H_2O \rightarrow$ Allyl alcohol + Acetic acid
Hydroformylation of allyl alcohol	Allyl alcohol + CO + $H_2 \rightarrow$ 4-hydroxybutanal
Hydrogenation of 4-hydroxybutanal	4-hydroxybutanal + $H_2 \rightarrow$ BDO

(iv) MAH Process Route	
Main Process Steps	Chemical Transformation
Esterification of MAH	MAH + Ethanol $\rightarrow$ Monoethyl maleate (MEM)
Purification of diethyl maleate	MEM + Ethanol $\rightarrow$ Diethyl maleate (DEM) + H_2O
Hydrogenation of diethyl maleate	DEM + $H_2 \rightarrow$ Diethyl succinate (DES) DES + $2H_2 \rightarrow$ GBL + 2 Ethanol GBL + 2 $H_2 \leftrightarrow$ BDO (reversible) BDO $\rightarrow$ THF + H_2O

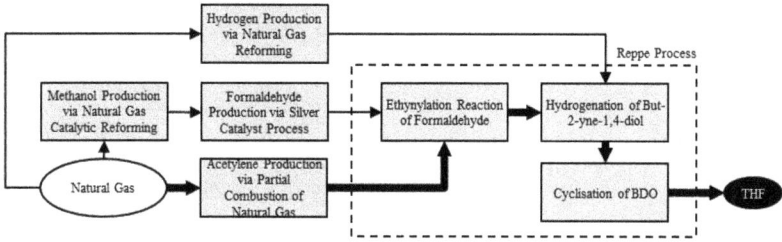

(a) Reppe process route

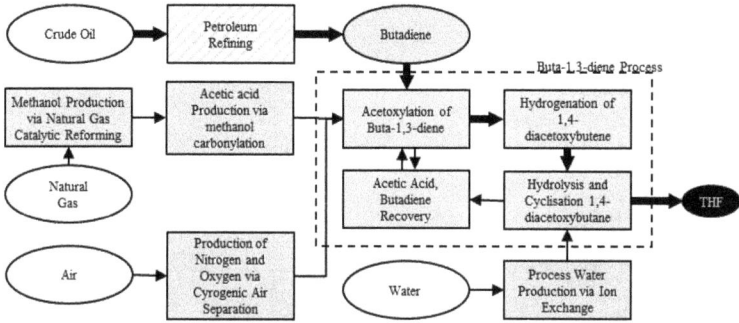

(b) Butadiene process route

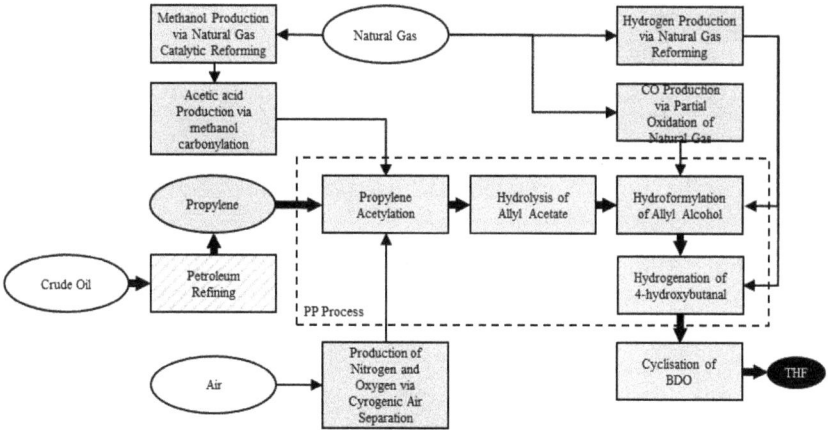

(c) Propylene process route

Figure 2. Life cycle stages involved in THF production: (a) Reppe process route; (b) Butadiene process route; (c) Propylene process route; (d) MAH process route.

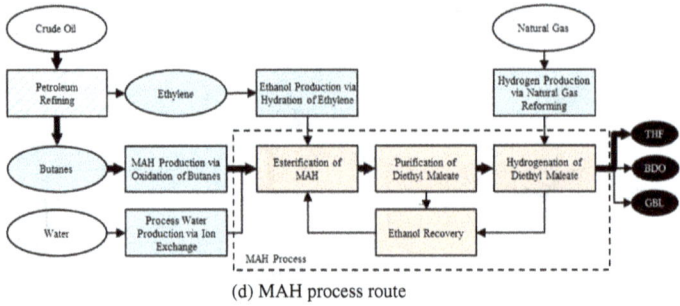

(d) MAH process route

Figure 2.   (*Continued*).

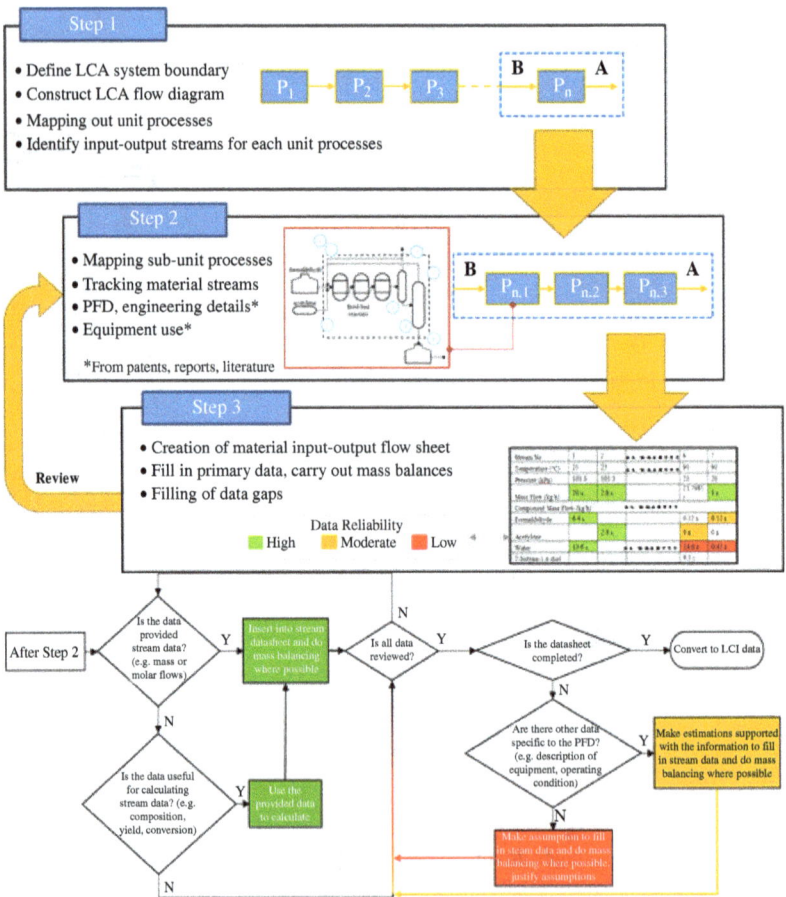

Figure 3.   Process-based approach and flowchart for obtaining material input–output LCI data.

Table 2. Data reliability according to *high* → *moderate* → *low*.

Levels	Examples
■ Highly reliable (green)	Calculation of the output stream of a reactor using conversion and selectivity.
	Calculation of stream data using weight or molar composition.
■ Moderately reliable (orange)	Using the top temperature of a separation column and boiling points to determine the gaseous and liquid chemical components.
	Using the middle value of a recommended range of values.
■ Less reliable (red)	Assuming a stoichiometric amount of reactant as input to a reactor due to the lack of information on the specific amount.
	No information to explain for changes in component mass flows after doing mass balance.

2.2.1. *Steps 1–3: example for route (i) Reppe*

In *Step 1*, the main life cycle production stages can be laid out after the LCA system boundary and scope has been clearly defined. *Step 2* proceeds to source for available information relating to the related production stages. According to information extracted from reports and/or patents, unit process boundaries are mapped out to identify the inputs, products, emissions, and intermediates. Figure 4 illustrates this important step for the Reppe process, accompanied by the generation of material flow streams (Tables 3 and 4).

The category values are weighted to calculate the overall EI value ($EI^r = \sum_n \sum_i w_i EI^r_{i,n}$), and by default $w_i = 0.2$ is used. Thereafter, the EI value is used in the following regression equation to derive the energy consumption estimate in terms of MJ/kg product ($E^{pred} = 57.9$ $EI - 1.8$).

2.2.2. *Energy estimations and corresponding greenhouse gas emissions*

Energy consumption is the major cost contributor to process operations as well as a major contribution to various environmental burdens.

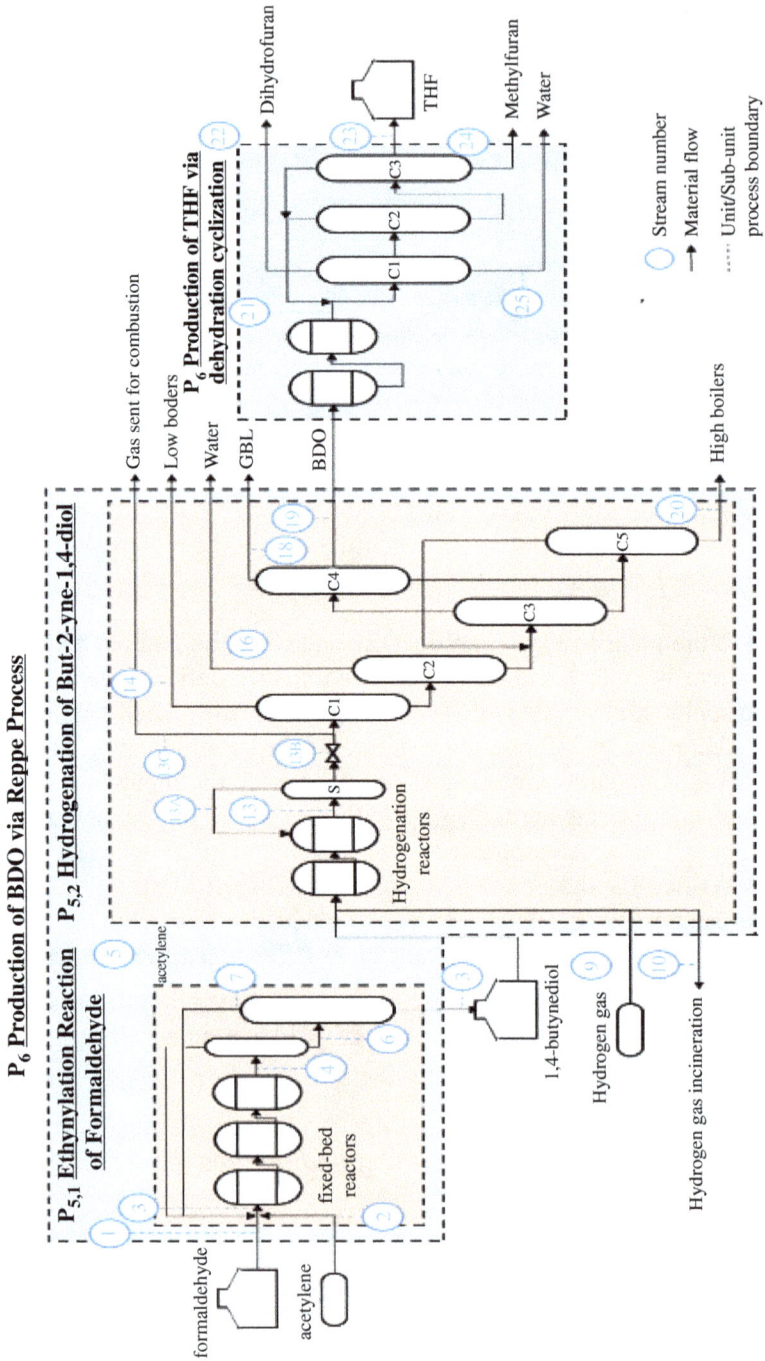

Figure 4.   Process flow diagrams for the sub-units of the Reppe process.

Table 3. Datasheet for $P_{5,1}$ ethynylation of formaldehyde sub-unit process. Letter subscripts denote the chronological steps in the construction of the datasheet.

Stream No.	1	2	3	4	5	6	7	8
Mass Flow (kg/h)	20_A	2.8_A	23.95_E	23.95_E	$1.5 \times 10^{-3}{}_C$	23.8_C	1_A	22.8_C
(a) Component Mass Flow (kg/h)								
Formaldehyde	6.4_A		6.52_E	0.12_F		0.12_D	0.12_E	—
Acetylene	—	2.8_A	2.95_B	0.15_C	$1.5 \times 10^{-3}{}_C$	0_B	0_B	0_B
Water	13.6_A	—	14.5_E	0.12_F	—	14.0_D	0.45_E	13.5_E
2-butyne-1,4-diol	—	—	—	9.3_C	—	9.3_C	—	9.3_E
Propynol	—	—	0.31_E	0.12_F	—	0.31_D	0.31_E	—
Methanol	—	—	0.12_E	0.12_F	—	0.12_D	0.12_E	—
Other organic compounds	—	—		0.12_F	—	0.0024_D	—	0.0024_E
(b) Component Molar Flow (kmol/h)								
Formaldehyde	0.21_A	—	—	—		—	—	—
Acetylene		0.11_A	0.11_B	0.0057_C	$5.4 \times 10^{-5}{}_C$	0_B		
Water	0.75_A	—	—			—	—	—
2-butyne-1,4-diol		—	—	0.11_C				

Large amounts of energy are used in driving mechanical equipment such as pumps, agitators, and process variable adjustors to facilitate the chemical transformation process. Pharmaceutical or chemical process industries are one of the sectors with a considerable share in energy consumption and corresponding greenhouse gas (GHG) emissions [1, 2]. One of the main challenges in obtaining the appropriate LCI data is in estimating the use of energy and its corresponding GHG emissions [5]. Unavailable information on energy use in various production stages was estimated using the methodology stated by Bumann *et al.* [6]. The methodology requires a simplified model of a process (illustrated in Figure 5), accompanied by a list of input parameters (Table 5) [6].

Table 4.   Summary of steps to complete the $P_{5,1}$ ethynylation of formaldehyde sub-unit process datasheet.

Step	Description
A	Direct transfer and calculation of stream data.
B	The absence of acetylene in Stream 6 was assumed as it has the lowest normal boiling point amongst species chemicals and has low solubility in a mixture of alcohols and aldehydes given its non-polar nature. Acetylene in Stream 3 was calculated from the sum of Stream 2 and 99% of Stream 4 as acetylene is recycled.
C	A conversion of 95% of formaldehyde and acetylene was used and Stream 5 is pure acetylene. Overall mass balance for Stream 5 and Stream 6.
D	Stream 6 was specified using the composition provided except for but-2-yne-1,4-diol. H_2O was used to balance Stream 6.
E	Based on the description of the separator, formaldehyde, methanol, propynol, and a portion of water are directed from Stream 6 to Stream 7. Stream 7 and 8 were specified using mass balance. Subsequently, Stream 3 was specified using mass balance.
F	Stream 4 was specified by mass balance from Stream 6.

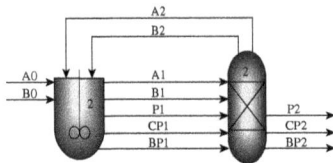

Fig.1. Simplified model of a chemical process with reactor (1) and separator (2). A and B are reactants, P is the desired product, CP is a coupled product that can not be avoided, BP is a by-product that is formed when the selectivity is not 100% The different numbers indicate stream labels.

Original scaling of the category inputs into category values between 0 and 1.

Category input	Category value	Input range
1) C_{H_2O} [mol/mol]	$EI_1 = 0$, no water present $EI_1 = 0.5$, water present $EI_1 = 1$, water to distill	Discrete scale
2) C_P [mol/mol]	$EI_2 = 1 - (1/2)\log_5(100.c_p)$	[0.25, 0.01]
3) ΔT_{Pb} [°C]	$EI_3 = 1 - (1/2)\log_5(\Delta T_{bp}/5))$	[20, 5]
4) $MLI_{CP,BP}$ [kg/kg]	$EI_4 = (1/2)(\log_{10} MLI_{CP,BP} + 1)$	[0.1, 10]
5) $\Delta_r H^\ominus > 0$ [kJ/mol] or $\Delta_r H^\ominus > 0$ and $T_R < 200$ °C Energy benefit: $\Delta_r H^\ominus < 0$ and $T_R > 200$ °C	$EI_5 = (\lvert\Delta_r H^\ominus\rvert - 100)/200$ $EI_5 = (\lvert\Delta_r H^\ominus\rvert - 100)/200$	[100, 300] [100, 300]

Figure 5.   Simplified model for a unit process and equations used to compute the five category values.

3. Results and Discussion

The LCA production, accompanied by the life cycle data generated, for the four process routes were modelled using SimaPro to generate the environmental impacts of energy use (in MJ) and global warming potential or GWP (kg CO_2-eq). The results are presented in Figures 6(a) and (b).

Table 5. List of input parameters required for the estimation of energy consumption.

c_{H2O}	Concentration of water at reactor outlet
c_p	Concentration of product at reactor outlet
ΔT_{bp}	Smallest absolute difference between the boiling point of the product and another substance that has to be separated from this product
μ_2^{BP}	Mass flow of by-products at separator outlet
μ_2^{CP}	Mass flow of co-products at separator outlet
μ_2^{P}	Mass flow of desired product at separator outlet
ΔH_{rxn}	Standard enthalpy change of reaction
T_R	Reactor temperature

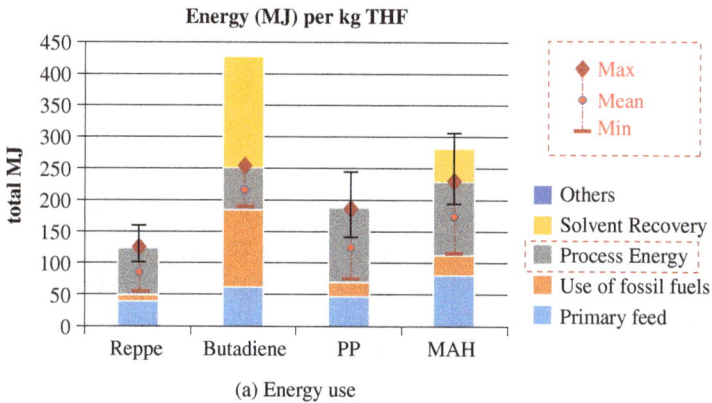

(a) Energy use

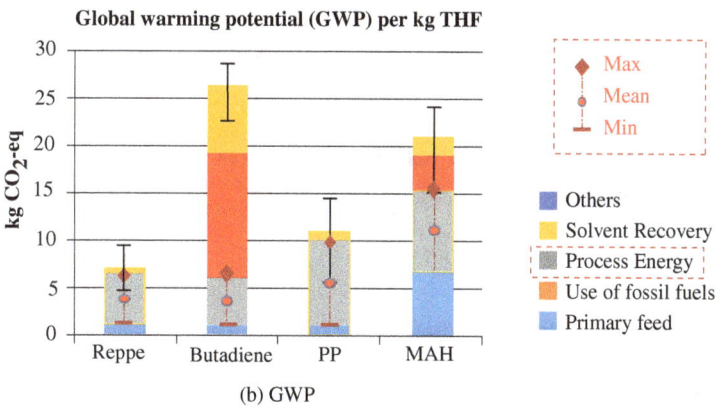

(b) GWP

Figure 6. Results: (a) Energy use and (b) GWP per kg THF: (a) Energy use; (b) GWP.

Figure 6(a) shows that the biggest variations in process energy come from cradle-to-gate industrial production routes of PP and MAH for the manufacture of THF. The range of MJ per kg can be –55 to 199 for the production of THF from PP and –110 to 250 for the production of THF from MAH. However, the highest (total mean) energy use was demonstrated by the butadiene route for producing THF at 430.2 MJ/kg and the Reppe process, which is the most common route, resulted in an average score of 123.4 MJ/kg THF. It should be highlighted that LCA results may vary significantly if there are any changes made in process operations requiring different equipment and associated resource utilization [7]. The energy consumption values required for LCA modeling were estimated since the exact operating conditions (e.g. separating columns, actual operating conditions such as pressure and temperature) for THF manufacturing were publicly unavailable [3, 4]. The MJ per kg THF values illustrated in Figure 6(b) give the possible minimum, mean, and maximum values as the result of equipment and some operating conditions used in the production processes.

Several similar MJ result trends (Figure 6(a)) can be deduced from the GWP results (Figure 6(b)) — the biggest variation of GWP results corresponds to the cradle-to-gate production routes of PP and MAH used for THF manufacture. Altogether, the largest portion (max values) of CO_2-eq emissions from butadiene and MAH process routes — a combination of primary feed, process energy, solvent recovery, etc. — are found to be from the upstream unit processes involved in the manufacturing steps involved in producing THF. Additionally, the butadiene process utilizes acetic acid and nitrogen which contributes (mean estimations of) 2.75 and 3.89 kg CO_2-eq/kg THF, respectively. As for the fourth production route, where chemical feedstock MAH is derived from the oxidation of butanes, the resulting max total is 6.72 kg CO_2-eq/kg THF.

4. Concluding Remarks: Data Sources and Uncertainties

All input–output data are apportioned to accomplish the product of interest in LCA applications, expressed quantitatively as a functional unit. However, despite the commonly misunderstood concept from (non-LCA) scientists, LCA is not a "black box" concept that can "instantaneously" yield environmental results on its own without the support of much needed information. Inadequate or poor quality data leads to inappropriate LCI which in turn produces unreliable LCA results.

Primary data — measured and gathered directly from the manufacturers or chemical industries — are often considered to be highly reliable for performing LCA. In the approach described in Figure 3, the accomplishment of *Step 2* requires appropriate manufacturing information sourced from patents, reports, literature, and most importantly, engineering-/process-based details. The results are expected to vary with different applications of LCI data selection based on the steps suggested.

Reliable LCA results depend on the availability of more detailed technical or process engineering information available to support the construction of the LCA model and its designs. Knowledge on chemical transformation will add to the credibility of data generation for use as life cycle inventory [8, 9]. Comprehensive or complex LCA modeling encompassing chemistry transformation and process-based knowledge will demand substantial time and effort from a group of chemical engineers and chemists [10]. However, such information is commonly not readily available for a complex chemical or pharmaceutical product synthesized from several different compounds [11].

As LCA evolves to serve different levels of sustainability objectives, it would prove worthy to exercise precaution and include extra steps in data selection [12, 13]. Figure 7 illustrates a framework describing the adequacy and completeness of life cycle inventory requirements applied for complex chemicals. Along with its growing use and advancement in

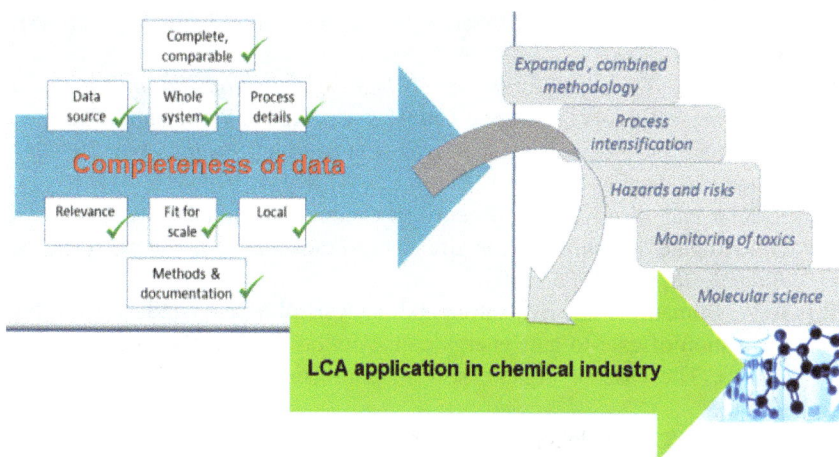

Figure 7. Framework: data completeness for LCA application involving complex chemical synthesis.

the industry, selecting adequate LCI of appropriate quality for LCA has been an ongoing challenge for many years [5, 7, 14]. This ongoing challenge is especially prevalent in the chemistry community, as echoed in the words of Jiménez-González *et al.* [15, p. 1489]:

> *it can be challenging to acquire the information needed to assess the resource and energy consumption and emissions generated for all phases over the life of a [chemical] product ...*

In a noteworthy piece of work, Kleinekorte *et al.* [16] presented the estimation methods for closing data gaps with the integration of LCA into process design. In their analysis, process designs are associated with the conceptual engineering of a chemical plant (e.g. the choice of equipment and process settings). Finally, as a necessary step toward proper management of LCA data requirements and models, Igos *et al.* [17] called for a more detailed review of uncertainty assessment in LCA.

References

[1]   D. Cespi, E. S. Beach, T. E. Swarr, F. Passarini, I. Vassura, *et al.*, Life cycle inventory improvement in the pharmaceutical sector: Assessment of the sustainability combining PMI and LCA tools. *Green Chem.* **17**, 3390–3400 (2015).

[2]   C. Jiménez-González, S. Kim, and M. Overcash, Methodology for developing gate-to-gate Life cycle inventory information. *Int. J. Life Cycle Assess.* **5**, 153–159 (2000).

[3]   H. Müller. Tetrahydrofuran. In: *Ullmann's encyclopedia of industrial chemistry*. (Wiley-VCH Verlag GmbH & Co. KgaA, Weinheim, 2012).

[4]   P. Dutia, *1,4-Butanediol: A Techno-Commercial Profile*. pp. 205–211 (Chemical Weekly, 2011).

[5]   S. Kim and M. Overcash, Energy in chemical manufacturing processes: Gate-to-gate information for life cycle assessment. *J. Chem. Tech. & Biotech.* **78**, 995–1005 (2003).

[6]   A. A. Bumann *et al.*, Evaluation and analysis of a proxy indicator for the estimation of gate-to-gate energy consumption in the early process design phases: The case of organic solvent production. *Energ.* **35**, 2407–2418 (2010).

[7]   C. Rérat, S. Papadokonstantakis, and K. Hungerbühler, Estimation and analysis of energy utilities consumption in batch chemical industry through thermal losses modeling. *Ind & Eng Chem Res.* **51**, 10416–10432 (2012).

[8] C. Jiménez-González and D. J. C. Constable, *Green Chemistry and Engineering: A Practical Design Approach* (John Wiley & Sons Inc., New Jersey, 2011).

[9] R. Hischier, S. Hellweg, C. Capello, and A. Primas, Establishing life cycle inventories of chemicals based on differing data availability. *Int. J. Life Cycle Assess.* **10**, 59–67 (2005).

[10] G. Reuss, W. Disteldorf, A. O. Gamer, and A. Hilt, *Formaldehyde: Ullmann's Encyclopedia of Industrial Chemistry* (Wiley-VCH Verlag GmbH & Co. KGaA, Hoboken, NJ, USA, 2000).

[11] G. Wernet, S. Conradt, H. P. Isenring, C. Jiménez-González, and K. Hungerbühler, Life cycle assessment of fine chemical production: A case study of pharmaceutical synthesis. *Int. J. Life Cycle Assess.* **15**, 294–303 (2010).

[12] H. H. Khoo, V. Isoni, and P. N. Sharratt, LCI data selection criteria for a multidisciplinary research team: LCA applied to solvents and chemicals. *Sustain. Prod. & Consump.* **16**, 68–87 (2018).

[13] W. Wei, P. Larrey-Lassalle, T. Faure, N. Dumoulin, P. Roux, and J.-D. Mathias, Using the reliability theory for assessing the decision confidence probability for comparative life cycle assessments. *Environ. Sci. Technol.* **50**, 2272–2280 (2016).

[14] G. Geiser, T. B. Hofstetter, and K. Hungerbühler, Production of fine and specialty chemicals: Procedure for the estimation of LCIs. *Int. J. Life Cycle Assess.* **9**, 101–113 (2004).

[15] C. Jiménez-González, D. J. C. Constable, and C. S. Ponder, Evaluating the "Greenness" of chemical processes and products in the pharmaceutical industry — A green metrics primer. *Chem. Soc. Rev.* **41**, 1485–1498 (2012).

[16] J. Kleinekorte, L. Fleitmann, M. Bachmann, *et al.*, Life cycle assessment for the design of chemical processes, products, and supply chains. *Annual Rev. Chem. Biomolecular Eng.* **11**, 203–233 (2020).

[17] E. Igos, E. Benetto, R. Meyer, P. Baustert, and B. Othoniel, How to treat uncertainties in life cycle assessment studies? *Int. J. Life Cycle Assess.* **24**(4), 794–807 (2010).

Chapter 3

Feasibility of CO$_2$ Capture and Utilization: From the LCA Perspective

**Albertus D. HANDOKO[*,‡], Hsien H. KHOO[†,§],
Iskandar HALIM[†,¶] and Karthick MURUGAPPAN[†,‖]**

*Institute of Materials Research and Engineering (IMRE),
Agency for Science, Technology & Research (A*STAR),
2 Fusionopolis Way, Innovis, Singapore 138634*

†*Institute of Chemical and Engineering Sciences (ICES), Agency for
Science, Technology & Research (A*STAR), Singapore 627833*

‡*adhandoko@imre.a-star.edu.sg*

§*khoo_hsien_hui@ices.a-star.edu.sg*

¶*Iskandar_Halim@ices.a-star.edu.sg*

‖*murugappan_karthick@ices.a-star.edu.sg*

Various CO$_2$ Capture and Utilization (CCU) methods have recently received a great deal of attention as potential solutions to reduce CO$_2$ emissions from major industrial emitters. In this work, we investigate the theoretical models of two CO$_2$-to-ethylene via electrochemical conversion methods for several small-scale (1 kg output) and large-scale (1,000-ton output) scenarios. The scope is to compare the net CO$_2$ emissions for both and highlight the technological challenges faced in the transition from lab to larger scales. The work outcome highlights that most CO$_2$ utilization technologies are still in their early development stage and should be carefully evaluated via a holistic life cycle perspective. Discussions on LCA modeling issues faced for various emerging CCU options are also addressed.

1. Introduction

Carbon capture and utilization as well as carbon capture and storage (CCU and CCS) are important emissions reduction technologies that can be applied across the energy system. Arguably, CCU represents a step forward from CCS, as it allows the captured CO_2 to be valorized, repurposed, or reutilized. CCU offers the possibility of carbon recycling, where CO_2 is converted into fuel or chemical precursor moieties that can be fed back into the production cycle. Various CCU methods have recently received a great deal of attention as potential solutions to reduce CO_2 emissions from major industrial emitters [1, 2]. In this context, the research of utilizing CO_2 as resources for conversion into chemical feedstock, fuels, or commodity products has piqued the interest of academia and industry alike [3, 4].

A schematic of the CO_2 valorization process along with the potential products that can be used as fuels or chemical feedstock is illustrated in Figure 1. Electrocatalysis is probably one of the most versatile CO_2 valorization approaches, with a clear pathway into industrial-scale production. Unlike thermal catalysis, the operation of electrocatalytic CO_2 reduction (CO_2RR) unit is simpler. CO_2RR can be operated on-demand at an adaptable scale, depending on the real-time electricity cost or demand with minimum energy penalty [5]. Lower energy cost driven by a significant rise in renewable electricity is certainly helpful in achieving higher CO_2 abatement target and financial viability. However, as the development of CO_2RR is still in the infancy stage, there is currently limited understanding of its net energy CO_2 emission balance, especially at a much larger scale [6].

In this chapter, we investigate CO_2RR from the Life cycle assessment (LCA) perspective. Specifically, we show that although it is relatively easy to demonstrate CO_2 abatement at a smaller scale, the net CO_2 balance at a larger scale depends greatly on the proper selection of the energy sources.

1.1. *CO_2 conversion processes*

After the capture of CO_2 from various sources, the next challenge is converting CO_2 into value-added products. The electrochemical CO_2 conversion method is explored in this study as this approach can

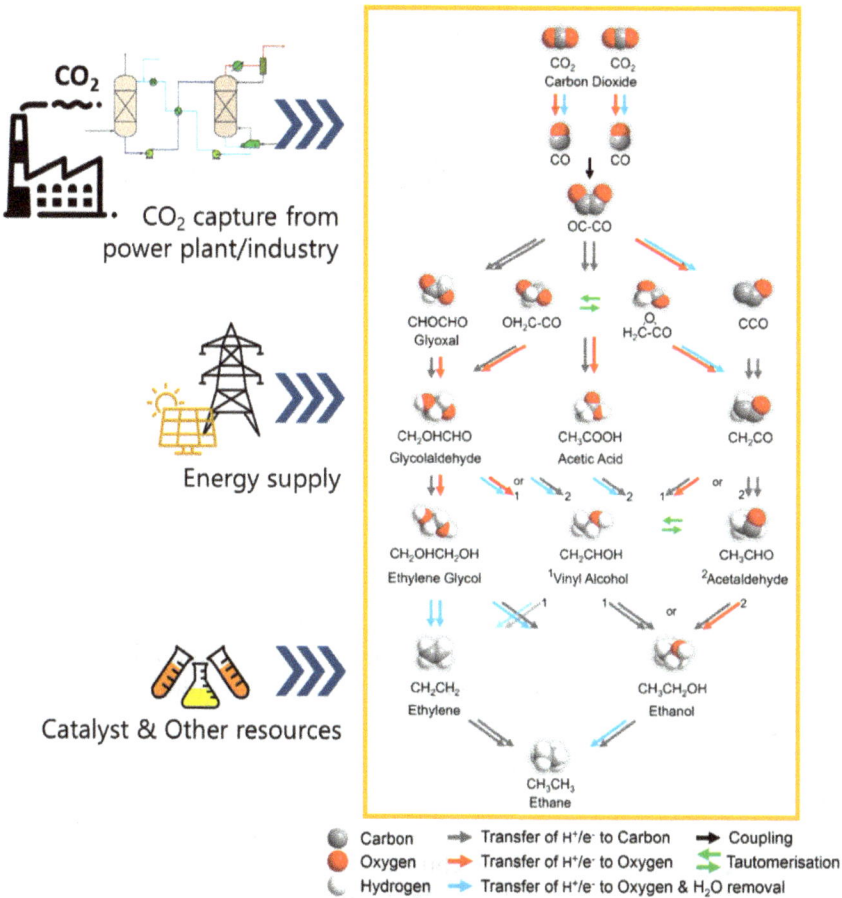

Figure 1.   Schematic diagram of possible schemes and resources input for CCU.
Source: Adapted from Ref. [12].

achieve a reasonable turnover at ambient temperature and pressure along with milder reaction conditions compared to thermochemistry [12, 13]. Additionally, upscaling can be achieved by introducing flow chemistry, increasing pressure, or increasing electrolyte concentration. More importantly, the electrochemical approach is readily compatible for coupling with renewable energy sources. With scalability and flexibility in its implementation, electrocatalysis can play a central role in CCU applications. One route for CO$_2$ electro-conversion is via the reduction of CO$_2$ and H$_2$O in an aqueous electrolyte to produce ethylene (C$_2$H$_4$):

$$2CO_2 + 2H_2O \rightarrow C_2H_4 + 3O_2 \qquad (1)$$

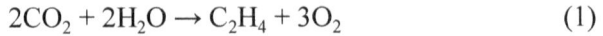

Based on (1), the thermodynamic energy required to obtain one mole of C_2H_4 can be calculated as

$$[68.2+0] - [2(-237.14) + 2(-394.39)] = 1331.26 \text{ kJ/mol} \qquad (2)$$

where the Gibbs free energy of formation values for C_2H_4, H_2O, CO_2, and O_2 are 68.2, –237.1, –394.4, and 0 kJ/mol, respectively. Therefore, the energy required for CO_2 conversion to ethylene in an electrochemical setup is estimated according to

$$1331.26 \text{ kJ/mol} \times \frac{100\%}{EE} \qquad (3)$$

Energy efficiency (EE) = Faradaic efficiency (FE) × voltage efficiency (VE).

Different electrolyzer designs or thermodynamics effects impose inherent constraints on the overall electrolytic performance. Other factors such as kinetics, power dissipation, and mass transport limitations depend on the operating conditions and tend to increase the required cell voltage [14, 15].

While CO_2RR has been shown to have great potential, Zimmermann and Schomäcker [6] cautioned that most CO_2 utilization technologies are still in their infancy and should be carefully evaluated. Nyári *et al.* [16], on the other hand, reported that techno-economic barriers may be encountered while scaling up the electrolyzer units and emphasized on issues relating to economic feasibility. Additionally, Jouny *et al.* [17] stated that the prospects for the large-scale production of chemicals via electrochemical reduction of CO_2 deserve further investigation.

2. Life Cycle Assessment Approach

LCA is a scientific and technical tool used to appraise the net CO_2 emission (or consumption) for the assessment of CCU options. The LCA approach accounts for additional greenhouse gas (GHG) emissions that may arise from energy generation and other indirect resources and therefore offers a system-level perspective. In the pursuit of ensuring the benefit of CCU, a few applications of LCA in the area of CO_2 valorization have been carried out. An example was given by von der Assen and

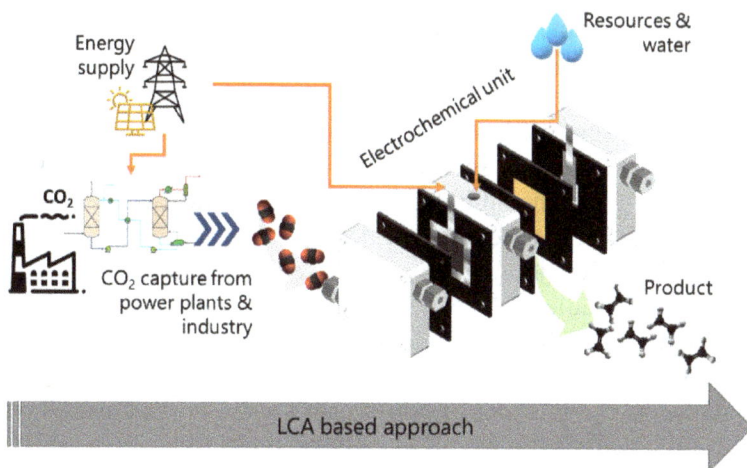

Figure 2. Overall LCA approach for the evaluation of CCU to produce commodity chemicals.

Source: Adapted from Ref. [5].

Bardow [18] on the LCA of polyols for polyurethane production using CO_2 as feedstock. Their production system was based on an industrial pilot plant. Another noteworthy study on an integrated review of catalysis and LCA for CO_2 conversion was reported by Artz *et al.* [19].

Sternberg *et al.* [20] presented a comprehensive LCA comparison of the production of formic acid, carbon monoxide, methanol, and methane from CO_2. For an overall sustainability assessment of CCU options, the production chain from CO_2 source to products, including capture technologies, energy, and resource requirements and their associated GHG emissions is ideally considered from the source of CO_2 pre-processing stages and delivery to the conversion facility to produce the final desired product [21]. Figure 2 illustrates an overall LCA approach for CCU aligned with the demand for commodity chemical feedstock for the chemical industry.

2.1. Life cycle assessment goal and scope: preliminary assessment of CO_2RR

As one of the most valuable olefin products of petrochemical industries worldwide, the case study focuses on CO_2-to-ethylene via electrochemical conversion [5, 12, 13]. While processes for CO_2 capture and CO_2RR are currently being intensively investigated, limited studies have addressed the total life cycle of CO_2 of both options combined. In this work, we investigate the theoretical models of two CO_2-to-ethylene via electro-chemical conversion methods for several small-scale (1 kg output) and large-scale (1-ton output) scenarios. The scope is to compare the net CO_2 emissions for both and highlight the technological challenges faced in the transition from lab to larger scales.

For a simplified LCA approach, the system boundary starts with sources of CO_2 (cradle) and next, the transfer of the gas to the conversion pathway and ends with the product of interest (gate). In lab-scale investi-gations, CO_2 sources are typically sourced from gas cylinders [12] and on larger scales, from industries or power plants [18, 21]. The overall analy-sis of routes that undertake CCU options must consider a life cycle approach of all the processes involved in order to determine if additional CO_2 emission occurs beyond the amount of CO_2 utilized. The scope of the LCA approach only considers CO_2 emissions since other types of pollut-ants are not reported in this specific case study. Transportation of CO_2 feedstock and catalyst use is also omitted.

2.1.1. *Small-scale setup: 1 kg ethylene output*

For the small-scale case study of CO_2RR, the following conditions are considered/assumed:

- The energy input options are from waste heat, solar power, and natural gas combustion.
- CO_2 is supplied from a gas tank, and therefore, CO_2 capture from power plants or other sources is not considered.
- Different amounts of CO_2 are emitted from different energy sources (waste heat, natural gas, and solar power).
- Mass balances for CO_2 amount to produce 1 kg of ethylene are calcu-lated according to total EE estimates of 80%, 70%, and 60%, respec-tively [15].
- Energy requirement for water supply is negligible.
- Net CO_2 utilized will be calculated.

The production, replenishment, and treatment of CO_2RR catalysts are not taken into consideration. The scenarios investigated for 1 kg of CO_2-to-ethylene conversion, including the total energy efficiencies and energy

Table 1. Summary of the small-scale (1 kg output) CO_2RR-to-ethylene conversion scenarios.

Scenario	Energy Providers	EE (%)	Energy Input (kWh)	CO₂ Input (kg)	Water Use (kg)	Separation/ Distillation (kWh)
I	WH	80	16.5	3.93	1.61	4.2
II	SR					
III	NG					
IV	WH	70	18.9	4.49	1.84	4.8
V	SR					
VI	NG					
VII	WH	60	22.0	5.24	2.14	5.6
VIII	SR					
IX	NG					

Note: EE = Energy efficiency; WH = Waste heat; SR = Solar power; NG = Natural gas.

requirements, are compiled in Table 1. All the input data are calculated according to Equations (1) and (2).

2.1.2. *Large-scale model: 1,000-ton ethylene output*

Despite successful process concepts carried out on a lab or bench scale, questions pertaining to the technological feasibilities of large-scale commercial applications remain [3, 6, 16]. These questions warrant further exploration into the potential net CO_2 emissions generated for large-scale scenarios. A second investigation is therefore carried out for the three proposed large-scale scenarios. Post-combustion carbon capture from flue gas is already in place for different industrial applications [24, 25], suggesting the feasibility of our large-scale proposition. We proposed the following three large-scale scenarios:

- 1,000-ton CO_2RR-to-ethylene plant located next to a petroleum refinery;
- 1,000-ton CO_2RR-to-ethylene plant located next to a waste-to-energy (WTE) plant and a solar plant;
- 1,000-ton CO_2RR-to-ethylene plant located next to a natural gas combined cycle (NGCC) power plant.

Detailed operating condition assumptions for the scenarios, categorized based on their primary energy source are as follows:

Organic Rankine Cycle (ORC) system harnessing waste heat

- A large-scale CO_2RR-to-ethylene plant is located next to a refinery.
- CO_2 supply is captured from the refinery flue gas with a solvent-based carbon capture technology [24].
- ORC system supplies part of the energy required for electrolysis [10, 11].
- Any additional power required for CO_2 capture, electrolysis, water pump, and final product distillation is supplied from the grid, with the power generation assumed to come from liquefied natural gas (LNG) combustion.

Solar power and Waste-to-Energy (WTE) plant

- A large-scale CO_2RR-to-ethylene plant is located next to a WTE system and a solar power plant.
- CO_2 supply is captured from the WTE facility with a solvent-based carbon capture technology [25].
- The power required for CO_2 capture will be self-supplied by the WTE plant from waste combustion. This renders energy requirement for CO_2 capture negligible in this case.
- Any additional power required for electrolysis, water pump, and final product distillation comes from solar-generated electrical power (forecasted power generation of solar is *ca.* 2,400 GWh/year).

NGCC power plant

- The large-scale CO_2RR model is located next to an NGCC power plant.
- CO_2 supply is captured from the NGCC power plant with a solvent-based carbon capture technology.
- Additional power supply required for CO_2 capture, electrolysis, water pump, and final product distillation is supplied from the NGCC power plant.

In all three cases, CO_2 transportation via piping is considered negligible. The scenarios for the large-scale cases of I to IX are compiled

Table 2.   A summary of the large-scale (1,000-ton output) CO₂RR-to-ethylene scenarios.

Scenario	Energy Providers	EE (%)	CO₂ Capture (MWh/t)	Energy Use (MWh)	CO₂ Supply (t)	Water (t) & Pump (MWh)	Separation/ Distillation Energy (MWh)
I	WH + power grid	80	1.2	16,509	3,927	1,608 (338)	625
II	SR + WTE		Neg				
III	NGCC power		2.4				
IV	WH + power grid	70	1.2	18,861	4,489	1,837 (386)	714
V	SR + WTE		Neg				
VI	NGCC power		2.4				
VII	WH + power grid	60	1.2	22,012	5,237	2,143 (450)	833
VIII	SR + WTE		Neg				
IX	NGCC power		2.4				

Note: EE = Energy efficiency; WH = Waste heat; WTE = Waste-to-Energy facility; SR = Solar power; NGCC = Natural gas combined cycle power plant; neg = Negligible.

in Table 2. Calculations for MJ requirements for CO₂ capture from the NGCC power plant are obtained from Khoo *et al.* [22]. Other types of power requirements for other operating systems (e.g. waste heat, separation/distillation) are estimated from Wei *et al.* [26] and Ren *et al.* [27].

3. Results and Discussion

The results for the small-scale and large-scale cases are displayed in Figures 3 and 4, respectively. Despite the favorable results (scenarios I, II, IV, V, VII, and VIII) for CO₂RR to 1 kg of ethylene shown in Figure 3, passionate deliberations on the technological feasibility of large-scale CO₂ conversion strategies beyond laboratory setups are still ongoing [28, 29]. One of the reasons lies in the fact that for experimental investigations carried out in lab or bench scales for producing chemicals or fuels from CO₂, the source of CO₂ feedstock conveniently comes from gas cylinders; therefore, any additional energy penalties associated with CO₂ capture and

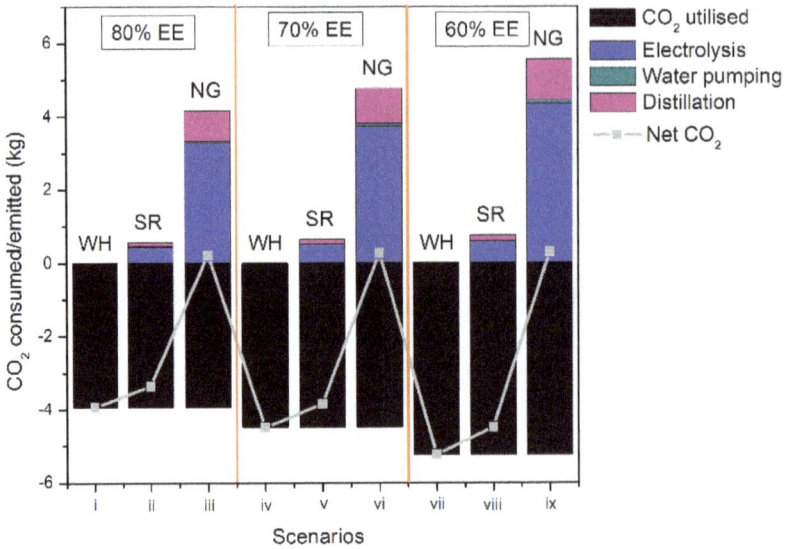

Figure 3.   Simulated results for small-scale CO_2RR to 1-kg ethylene.

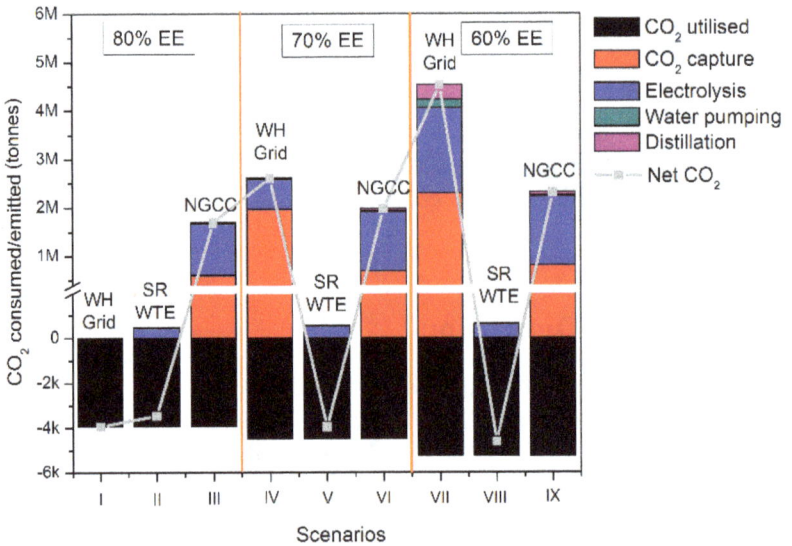

Figure 4.   Simulated results for large-scale CO_2RR-to-1,000-ton ethylene (M = million; k = kilo).

purification technologies or delivery will not be taken into account [16, 17, 23].

As displayed in Figure 4, the overall results of scenarios I, III, IV, VI, VII, and IX demonstrate that large-scale applications of CO_2 electrochemical conversion are unfeasible in these cases, as the amount of CO_2 utilized pales in comparison to the amount emitted from CO_2 capture and/or electrolysis. A large amount of energy input is required for the electrolysis plant (16,509–22,012 MWh for 1,000-ton C_2H_4 produced) with an associated net CO_2 emission of 1.69–4.53 million tons per 1,000-ton C_2H_4 produced. The significant increase in CO_2 is emission is due to the CO_2 capture step becoming significantly more energy-intensive as the CO_2 feed requirement vastly increases. Higher levels of associated CO_2 emissions are generated when non-renewable or unsustainable energy sources are used.

One of the key challenges for advancing CO_2RR lies in increasing the EE% of the electrochemical system, which is primarily hindered by high overpotential [3, 17]. As a distinct contrast to small-scale scenarios, a leap toward larger-scale industrial-sized employment comes with a new set of engineering and technological challenges [7, 16, 30].

4. Further Discussion: Life Cycle Assessment Modeling for CCU Applications

As CCU attracts much attention in the scientific arena [1, 3–5] for CO_2 emission mitigation, the potential benefits — or any unintended drawbacks — of CCU applications should be analyzed in a holistic method such as LCA [19]. In comprehensive investigations of any types of new technologies or emerging processes, LCA typically investigates various environmental impact categories over the entire life cycle of products or services [23]. For a comprehensive or more complete LCA investigation, various types of emissions (e.g. dust, volatile organic compounds, CO, SO_2, NO_X) leading to other environmental impact categories such as human toxicity, acidification, and eutrophication, will typically be considered [22, 31].

LCA studies may show varying results for CCU technologies due to specific operating parameters and assumptions applied in LCA models of assessment [22, 32]. One of the reasons lies in the different scope and definition of the "functional units" (that is, the defined quantity and quality of the product or service output) as well as the relative basis for which the environmental impacts are calculated and interpreted.

In order to address the variabilities or changes in parameters applied in LCA modeling of CCU technologies, uncertainty analysis can be applied as a measurement of the reliability of the model output toward the underlying decision process [33]. A recent study by Müller *et al.* [34] elaborates that the central part of CCU methods involves the capture and supply of CO_2 feedstock. The authors claim that GHG emission reductions by CCU depend strongly on the choice of the CO_2 source, because CO_2 sources differ in CO_2 concentration and the resulting energy demand for capture. Such factors can influence CCU results and selections. These differences can substantially impact the selection of environmentally beneficial CO_2 sources in industry and policymaking and even the perception of CCU in general. As an example, Dimitriou *et al.* [35] show that the net GHG emissions can vary among CO_2 sources, due to the different CO_2 concentrations, impurities, and available carbon capture methods. In order to avoid any pitfalls in LCA modeling, a more comprehensive systematic framework for LCA of CCU has been proposed by von der Assen *et al.* [36], including a thorough assessment of the environmental impacts beyond carbon and energy balance.

5. Concluding Remarks

This chapter elaborates and discusses why a life cycle perspective is mandatory to fully evaluate the net CO_2 avoidance potential of utilizing CO_2 as a feedstock for making products. A life cycle approach ensures that all additional CO_2 emissions from energy or resource use are taken into account [32, 33, 36].

In conclusion, a holistic evaluation using a complete LCA investigation is highly recommended. Cuéllar-Franca and Azapagic [23] demonstrated that in the investigation of CCU options aimed for greenhouse gas reduction, other environmental consequences such as acidification and toxicity potential also have to be carefully considered. Overall, this chapter highlights the importance of carefully assessing the feasibility of CCU effort, which can help us obtain the bigger picture via LCA and identifies existing scientific and engineering challenges that should be overcome in order to make CO_2 valorization a more promising technology for a sustainable future.

References

[1] A. Barthel, Y. Saih, M. Gimenez, J. D. A. Pelletier, F. E. Kühn, and V. D'Elia, Highly integrated CO_2 capture and conversion: Direct synthesis of cyclic carbonates from industrial flue gas. *Green Chem.* **18**, 3116–3123 (2016).

[2] D. T. Whipple and P. J. A. Kenis, Prospects of CO_2 utilization via direct heterogeneous electrochemical reduction. *J. Phy. Chem. Lett.* **1**, 3451–3458 (2010).

[3] A. S. Reis Machado and M. Nunes da Ponte, CO_2 capture and electrochemical conversion. *Current Opin. Green Sustain. Chem.* **11**, 86–90 (2018).

[4] R. Chauvy, N. Meunier, D. Thomas, and G. De Weireld, Selecting emerging CO_2 utilization products for short- to mid-term deployment. *Appl. Energ.* **236**, 662–680 (2019).

[5] K. Liu, W. A. Smith, and T. Burdyny, Introductory guide to assembling and operating gas diffusion electrodes for electrochemical CO_2 reduction. *ACS Energy Lett.* **4**(3), 639–643 (2019).

[6] A. W. Zimmermann and R. Schomäcker, Assessing early-stage CO_2 utilization technologies — comparing apples and oranges? *Energ. Technol.* **5**, 850–860 (2017).

[7] G. Centi, E. A. Quadrelli, and S. Perathoner, Catalysis for CO_2 conversion: A key technology for rapid introduction of renewable energy in the value chain of chemical industries. *Energ. Environ. Sci.* **6**, 1711–1731 (2013).

[8] J. B. Obi, State of art on ORC applications for waste heat recovery and micro-cogeneration for installations up to 100 kWe. *Energ. Procedia* **82**, 994–1001 (2015).

[9] B. Xu, D. Rathod, A. Yebi, Z. Filipi, S. Onori, and M. Hoffman, A comprehensive review of organic rankine cycle waste heat recovery systems in heavy-duty diesel engine applications. *Renew. Sustain. Energ. Rev.* **107**, 145–170 (2019).

[10] E. H. Wang, H. G. Zhang, B. Y. Fan, M. G. Ouyang, Y. Zhao, and Q. H. Mu, Study of working fluid selection of organic Rankine cycle (ORC) for engine waste heat recovery. *Energ.* **36**(5), 3406–3418 (2011).

[11] B. Saleh, G. Koglbauer, M. Wendland, and J. Fischer, Working fluids for low-temperature organic Rankine cycles. *Energ.* **32**(7), 1210–1221 (2007).

[12] A. D. Handoko, F. Wei, Jenndy, B. S. Yeo, and Z. W. Seh, Understanding heterogeneous electrocatalytic carbon dioxide reduction through operando techniques. *Nature Catalysis* **1**(12), 922–934 (2018).

[13] Z. W. Seh, J. Kibsgaard, C. F. Dickens, I. Chorkendorff, J. K. Nørskov, and T. F. Jaramillo, Combining theory and experiment in electrocatalysis: Insights into materials design. *Science* **355**(6321), eaad4998 (2017).

[14] D. Ren, Y. Deng, A. D. Handoko, C. S. Chen, S. Malkhandi, and B. S. Yeo, Selective electrochemical reduction of carbon dioxide to ethylene and ethanol on copper(I) oxide catalysts. *ACS Catalysis* **5**(5), 2814–2821 (2015).

[15] A. J. Martin, G.O. Larrazabal, and J. Perez-Ramirez, Towards sustainable fuels and chemicals through the electrochemical reduction of CO_2: Lessons from water electrolysis. *Green Chem.* **17**(12), 5114–5130 (2015).

[16] J. Nyári, M. Magdeldin, M. Larmi, M. Järvinen, and A. Santasalo-Aarnio, Techno-economic barriers of an industrial-scale methanol CCU-plant. *J. CO_2 Util.* **39**, 101166 (2020).

[17] M. Jouny, W. Luc, and F. Jiao, General techno-economic analysis of CO_2 electrolysis systems. *Ind. & Eng. Chem. Res.* **57**(6), 2165–2177 (2018).

[18] N. von der Assen and A. Bardow, Life cycle assessment of polyols for polyurethane production using CO_2 as feedstock: Insights from an industrial case study. *Green Chem.* **16**(6), 3272–3280 (2014).

[19] J. Artz, T. E. Müller, K. Thenert, J. Kleinekorte, R. Meys, A. Sternberg *et al.,* Sustainable conversion of carbon dioxide: An integrated review of catalysis and life cycle assessment. *Chem. Rev.* **118**, 434–504 (2018).

[20] A. Sternberg, C. M. Jens, and A. Bardow, Life cycle assessment of CO_2-based C1-chemicals. *Green Chem.* **19**(9), 2244–2259 (2017).

[21] H. H. Khoo, P. N. Sharratt, J. Bu, T. Y. Yeo, A. Borgna, J. G. Highfield *et al.,* Carbon capture and mineralization in Singapore: Preliminary environmental impacts and costs via LCA. *Ind. & Eng. Chem. Res.* **50**(19), 11350–11357 (2011).

[22] H. H. Khoo, I. Halim, and A. D. Handoko, LCA of electrochemical reduction of CO_2 to ethylene. *J. CO_2 Util.* **41**, 101229 (2020).

[23] R. M. Cuéllar-Franca and A. Azapagic, Carbon capture, storage and utilisation technologies: A critical analysis and comparison of their life cycle environmental impacts. *J. CO_2 Util.* **9**, 82–102 (2015).

[24] P. Bains *et al.,* CO_2 capture from the industry sector. *Prog. Energ. Combust. Sci.* **63**, 146–172 (2017).

[25] I. Durán *et al.,* Separation of CO_2 in a solid waste management incineration facility using activated carbon derived from pine sawdust. *Energies* **10**, 827–847 (2017).

[26] M. Wei *et al.,* Study on the integration of fluid catalytic cracking unit in refinery with solvent-based carbon capture through process simulation. *Fuel* **219**, 364–374 (2018).

[27] T. Ren, M. Patel, and K. Blok, Olefins from conventional and heavy feedstocks: Energy use in steam cracking and alternative processes. *Energ.* **31**(4), 425–451 (2006).

[28] Z. Yuan, M. R. Eden, and R. Gani, Toward the development and deployment of large-scale carbon dioxide capture and conversion processes. *Ind. Eng. Chem. Res.* **55**(12), 3383–3419 (2016).

[29] R. Chauvy and G. de Weireld, CO$_2$ utilization technologies in Europe: A short review. *Energ. Technol.* **8**, 2000627 (2020).

[30] J. M. Bonem, *Chemical Projects Scale Up: How to go from Laboratory to Commercial* (Elsevier Science, Amsterdam, Netherlands, 2018).

[31] R. Arvidsson, A.-M. Tillman, B. A. Sandén, M. Janssen, A. Nordelöf, D. Kushnir *et al.*, Environmental assessment of emerging technologies: Recommendations for prospective LCA. *J. Ind. Ecol.* **22**(6), 1286–1294 (2017).

[32] B. Anicic, P. Trop, and D. Goricanec, Comparison between two methods of methanol production from carbon dioxide. *Energ.* **77**, 279–289 (2014).

[33] S. Pfingsten, D. O. Broll, N. von der Asses, and A. Bardow, Second-order analytical uncertainty analysis in life cycle assessment. *Environ. Sci. Technol.* **51**, 13199–13204 (2017).

[34] L. J. Müller, A. Kätelhön, S. Bringezu, S. McCoy, S. Suh, R. Edwards *et al.*, The carbon footprint of the carbon feedstock CO$_2$. *Energ. & Environ. Sci.* **13**, 2979 (2020).

[35] I. Dimitriou, P. García-Gutiérrez, R. H. Elder, R. M. Cuéllar-Franca, A. Azapagic, and R. W. K. Allen, Carbon dioxide utilisation for production of transport fuels: Process and economic analysis. *Energ. Environ. Sci.* **8**, 1775–1789 (2015).

[36] N. von der Assen, J. Jung, and A. Bardow, Life-cycle assessment of carbon dioxide capture and utilization: Avoiding the pitfalls. *Energ. Environ. Sci.* **6**(9), 2721–2734 (2013).

Chapter 4

Life Cycle Assessment Strategies for Carbon Capture and Utilization Processes

Arnab DUTTA

Department of Chemical Engineering,
Birla Institute of Technology and Science (BITS),
Pilani, Hyderabad Campus, Jawahar Nagar, Kapra Mandal,
Hyderabad 500078, Telangana, India

arnabdutta@hyderabad.bits-pilani.ac.in

Carbon capture and utilization (CCU) has been perceived as a promising strategy to curb CO_2 emissions. To assess the environmental impact of any CCU process, a holistic approach is desirable. This necessitates the importance of life cycle assessment (LCA) studies to unveil the complete potential environmental performance of any CCU process. This chapter presents an overview of the CCU system and focuses on techniques that should be adopted to conduct an LCA study in the context of CCU systems. It lays emphasis on defining the objective and boundary for the CCU system, data acquisition using LCA software and databases, and finally interpretation of LCA results. It also presents innovative approaches that can be a part of the LCA methodology, such as assessing time-dependent global warming potential and accounting for both economic and societal impacts. Following this, the chapter also presents the application of LCA for certain CCU systems. The combination of process simulation, integration, and optimization with LCA will indeed be the future direction within the CCU-based research community.

1. Background

Carbon capture and utilization (CCU) is based on the concept of capturing CO_2 from point sources followed by its conversion into various valuable products [1, 2]. The successful implementation of a CCU-based process has the potential to reduce CO_2 emissions, along with the depletion of fossil resources. The objective of any CCU-based process is not only to achieve economic benefits but also to reduce environmental impacts. In this context, it is imperative that the reduction of environmental impacts via a CCU process calls for a careful assessment. The market potential of CO_2-derived products is one of the key factors in realizing the full reduction potential of CCU [3, 4]. The reduction potential also depends on the energetic efficiency of the CCU process including both CO_2 capture and CO_2 utilization processes. CO_2 capture in power plants via amine absorption needs a substantial amount of energy for solvent regeneration [5]. Often, high energetic co-reactants like hydrogen are needed to activate the chemically inert CO_2 and obtain value-added CO_2-based products [6, 7]. The process of obtaining these co-reactants is usually associated with high environmental impacts in the form of indirect CO_2 emissions and resource depletion. One cannot completely rule out the possibility of a CCU-based process to be even less sustainable as compared to its conventional counterpart.

Thus, in order to evaluate the actual net benefit (or unintentional negative impacts), a CCU-based technology requires a detailed environmental assessment. To assess the environmental sustainability of a process, various metrics have been proposed over the year [8–10]. One of the most popular methods applied for the evaluation of CCU is life cycle assessment (LCA) [11–13]. Before delving into the details of LCA for CCU systems, the concept of CCU is presented in the following section. This is followed by a detailed discussion on "know-how" related to the implementation of LCA for CCU systems in Section 3. LCA results pertaining to certain CCU systems are presented in Section 4. Finally, the concluding remarks are presented in Section 5.

2. Carbon Capture and Utilization System

Although various carbon capture and sequestration (CCS) technologies were seen as a means to mitigate climate change, after several years, CCS has been under scrutiny primarily owing to the uncertainties related to the CO_2 leakage from ecological storage systems. CCS is not a feasible option either due to the constraint on geological storage capacity or only offshore provision being available, which affects the overall process economics [14]. These concerns related to CCS have essentially led to CCU as an alternative strategic measure to curb CO_2 emissions. Unlike CCS, CCU results in several value-added products that can be sold, thus resulting in revenue generation. Also, the utilization of CO_2 as a feedstock can be perceived as the utilization of a waste product and compared to traditional petrochemical feedstocks, CO_2 can be regarded as a renewable resource. The main difference between CCS and CCU lies in the ultimate fate of the CO_2 captured. In CCS, CO_2 is captured and then it is subjected to long-term storage, whereas in the case of CCU, CO_2 is transformed into different value-added products [15].

2.1. *CO_2 capture technologies*

Irrespective of CCS or CCU, CO_2 capture is the first step toward CO_2 mitigation. In this context, different CO_2 capture options are briefly presented before discussing different CO_2 utilization options. The CO_2 capture options can be classified as post-conversion, pre-conversion, and oxy-fuel combustion [16]. Post-conversion capture involves separating CO_2 from the waste gas streams emitted from the combustion of fossil fuels. Chemical absorption using different solvents, solid sorbents-based adsorption, cryogenics, membrane-based separation, and pressure and vacuum swing adsorption are some of the most commonly used strategies for post-conversion CO_2 capture [17, 18]. Out of these, chemical absorption via monoethanolamine (MEA) solvent is the most commonly used separation technique [15] (e.g. applied in Chapter 3).

Pre-conversion capture refers to the capturing of CO_2 generated as a product in a conversion process [16, 19]. For example, in the ammonia production process, CO_2 co-produced with hydrogen during steam reforming must be removed before proceeding with the ammonia synthesis reaction. Likewise, in an integrated gasification combined cycle

(IGCC) power plant, CO_2 must be separated from hydrogen prior to the gasification process. This can be achieved using physical solvents such as selexol and rectisol [16, 19].

Oxy-fuel combustion can be applied exclusively to combustion processes like power generation, cement manufacturing, and iron and steel industries [15]. The fuel is burned with pure oxygen, thus resulting in flue gas with high CO_2 concentrations and free of nitrogen and its compounds. Although oxy-fuel combustion avoids the need of separating CO_2 from flue gas, it involves high energy demand of the air separation process for oxygen production. Alternatives to the oxy-fuel process include chemical looping combustion and chemical looping reforming, both of which include oxygen transfer selectively from an air reactor to a fuel combustor using a metal oxide [15].

2.2. *CO_2 utilization opportunities*

The captured CO_2 can be used as a commercial product either directly or after a chemical transformation. CO_2 can be directly used in the food and drink industry as well as in enhanced oil recovery. CO_2 can also be used as a feedstock to produce chemicals and fuels. The different available options for utilizing CO_2 are discussed next.

2.2.1. *Direct utilization of CO_2*

CO_2 is commonly used in the food and beverage industry as a carbonating agent, preservative, packaging gas, as a solvent for extraction of flavors, and in the decaffeination process. CO_2 also finds extensive application in the pharmaceutical industry either as a respiratory stimulant or as an intermediate in drug synthesis. However, most of these applications demand a high purity CO_2 stream.

2.2.2. *Enhanced oil recovery*

Enhanced oil recovery (EOR) is another application that involves the direct utilization of CO_2. In this case, crude oil is extracted from an oil field using CO_2 as an agent. EOR involves the extraction of unrecoverable oil trapped in the rocks by the injection of external agents like CO_2, natural gas, nitrogen, polyacrylamides, and surfactants into the reservoir. CO_2 is the preferred choice owing to its wide availability and low cost.

2.2.3. *CO_2 to value-added products (chemicals and fuels)*

Transformation of CO_2 to chemicals and fuels can be achieved via carboxylation reactions that involve cleavage of the C=O bonds to produce chemicals such as methane, methanol, syngas, urea, and formic acid [1]. CO_2 can also be used as a feedstock to produce fuels via the Fischer–Tropsch process [1]. However, chemicals and fuels usually have a limited time for CO_2 storage owing to their shorter life span.

2.2.4. *CO_2 to minerals*

CO_2 reacts with metal oxides like magnesium or calcium to form carbonates. This process is known as mineral carbonation. It encompasses a series of reactions that can take place either in a single or a multi-step process [14]. The single-step or direct carbonation process usually takes place under high pressure in either dry or aqueous media. Multi-step or indirect carbonation consists of three reactions. The first step involves the separation of the metal from the mineral matrix using an extracting agent like hydrochloric acid or molten salts followed by a series of hydration reactions resulting in metal hydroxide. CO_2 reacts with this metal hydroxide forming a carbonate. Mineral carbonation results in the formation of stable carbonates capable of storing CO_2 for long periods without the risk of CO_2 leakage.

3. Life Cycle Assessment for Carbon Capture Utilization Systems: Case Study Examples

3.1. *System boundary concepts for Carbon Capture Utilization*

A fair comparison basis for end products, such as methanol or other chemicals, should be clearly defined (e.g. volume of output, purity level, etc.) in any type of LCA case study. One can refer to Chapter 1 for the illustration of system boundary conditions for various types of products and Chapter 3 for how system boundary parameters affect CCU results. In addition to these descriptions, a few more examples are given in this section.

3.1.1. *Comparison of polymers from CO_2 vs. petrochemical polymers*

Polymers can be similar products with identical functions. However, their technical performance may differ owing to some properties of the material. As a result, this might lead to a higher demand of material per packaging unit for one polymer. As LCA is based on comparing products with similar quality, owing to higher demand of a material, one of the polymers may be environmentally less sustainable even though it might be preferable on a kilogram basis. In such a scenario, it is recommended to change the basis of comparison. Instead of comparing on a kilogram basis, we need to focus on the packaging unit, i.e. compare a packaging unit made of 1 kg of a petrochemical polymer with an equally performing packaging unit made out of 1.2 kg of a CO_2-based polymer [10].

3.1.2. *Inclusion of co-products in life cycle assessment of carbon capture utilization*

Integrated production of chemicals can be economically as well as environmentally favorable with respect to their individual counterparts. This is a well-known concept in the chemical industry and CCU can further advance the integration within the chemical industry. In an LCA study, we need to account for all products within the system boundaries and in the functional unit. The environmental impacts are then calculated for the entire system and related to all products, e.g. kg CO_2-eq emissions per x kg of product A and y kg of product B. In such cases, this is known as system expansion [20].

3.1.3. *Comparison of global warming potential impacts of CO_2-based and fossil-based methanol*

Let us take a simplified LCA-based comparison of methanol production from CO_2 with that from natural gas [10]. Figure 1 presents the flow diagrams for comparing methanol production from CO_2 and natural gas. The functional unit defined 1000 kg of methanol as the main product and 1273 kWh of electricity as the co-product. In this case, the global warming potential (GWP) for CCU (CO_2-derived methanol) and

153 kg CO$_2$ **418 kg CO$_2$**   **188 kg CO$_2$**   **1090 kg CO$_2$**   **745 kg CO$_2$**

| Power Plant | 1375 kg CO$_2$ | MeOH Synthesis | 188 kg H$_2$ | Renewable H2 via Electrolysis |

| Power Plant | | MeOH Synthesis |

1273 kWh   **1000 kg**   **1273 kWh**   **1000 kg**
Electricity   **MeOH**   **Electricity**   **MeOH**

(a) **CCU system**   (b) **non-CCU system**

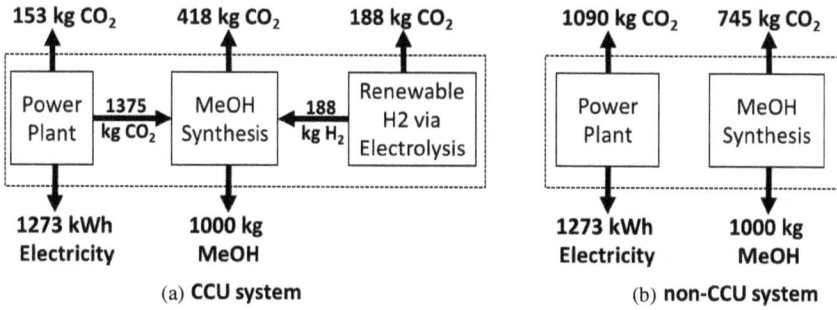

Figure 1. (a) CCU system comprising a power plant with CO$_2$ capture and methanol production from CO$_2$ and H$_2$ (H$_2$ obtained via electrolysis using wind power); (b) Non-CCU system comprising a power plant without CO$_2$ capture and traditional methanol synthesis via natural gas.

Source: Adapted from Ref. [10] with permission from The Royal Society of Chemistry.

non-CCU (fossil-derived methanol) process are GWP$_{CCU}$ = (153 + 418 + 188) = 759 kg CO$_2$-eq per functional unit and GWP$_{non-CCU}$ = (1090 + 745) = 1835 kg CO$_2$-eq per functional unit, respectively [10]. It is evident from this simple example that the CO$_2$-based methanol production is favorable, as it shows about 59% reduction in GWP impact as compared to conventional fossil-based production of methanol.

However, it is not always viable to take into account all products within the life cycle as a part of the functional unit. In this context, what we can alternatively do is calculate product-specific environmental impacts. One of the approaches here is to credit the by-products with CO$_2$ avoidance [21]. In the methanol production example, electricity is the by-product that can replace the conventional production of electricity, thus avoiding a certain amount of CO$_2$ relating to power production. This is credited by evaluating the CO$_2$ avoidance as (759–1090) = −331 kg CO$_2$-eq per 1000 kg of methanol. In this case, electricity is assigned the remaining emissions of 1090 kg CO$_2$-eq per 1273 kWh electricity, which is exactly the same as in the conventional power plant. Thus, this would not give any incentive for a power plant operator to adopt the CCU-based methanol production route. Now, if we reverse the viewpoint and look from the perspective of the power plant operator, methanol becomes the by-product. The CO$_2$ avoidance becomes (759–745) = 14 kg CO$_2$-eq per

1273 kWh electricity and the remaining emissions of 745 kg CO_2-eq per 1000 kg of methanol, which is the same as the fossil-based methanol route. In this alternative scenario, the methanol manufacturer has no environmental incentive to adopt CCU. It is clear that if we adopt the CO_2 avoidance approach, it will always credit the environmental benefits completely to one party and the other gets no incentive to switch to CCU.

In order to alleviate the above-mentioned issue, economic allocation [10] is another approach that can be used to calculate the product-specific environmental impacts. CO_2 is a valuable feedstock in CCU, thus greenhouse gas (GHG) emissions at the power plant can be allocated to two products, i.e. electricity and the CO_2 feedstock. Let us assume some price for both electricity and CO_2, say €50 per MWh of electricity and €60 per ton of CO_2, the power plants' revenues become $[1273*50/(1273*50 + 1375*60)] = 44\%$ from electricity generation and 56% from feedstock CO_2 production. This ratio can be used for the economic allocation of impacts [10]. Thereby, the individual products generate results of $[153*44\%] = 67$ kgCO_2-eq per 1273 kWh electricity and 86 kg CO_2-eq per 1375 kg feedstock of CO_2, i.e. $GWP(CO_{2,feed}) = 0.06$ kgCO_2-eq per kg $CO_{2,feed}$. The methanol then causes $(86 + 418 + 188) = 692$ kg CO_2-eq per 1000 kg methanol.

3.2. *Data-related issues for life cycle assessment of carbon capture utilization*

Comprehensive LCA calculations require a lot of data (refer to Chapter 1 again). Relevant LCA databases for chemical production are very much needed for comprehensive LCA investigations, e.g. covering a variety of products spanning from chemicals to power generating systems, transportation modes, recycling technologies, and waste treatment options [22]. With the use of datasets available, efficient metrics can be generated for aiding decision-making in CCU options. Prior to rigorous LCA evaluations, it is good practice to conduct a simplified version of LCA to have an estimate of the "best-case scenario". Table 1 presents some sample results about global warming and fossil resource depletion for certain feedstocks relevant to CCU. If the "best-case scenario" results in lower environmental impacts as compared to the existing conventional production, then it shows promise for further consideration. Otherwise, if negative environmental burdens are detected for the "best-case" options,

Table 1. Overview of global warming and fossil resource depletion impacts caused by the production of feedstocks frequently used in CO_2 chemistry per kg of feedstock.

Chemical	Global Warming Potential (kg CO_2-eq/kg)	Fossil Resource Depletion (kg oil-eq/kg)
CO_2 from coal-based power plant	(−0.7) − (+0.1)	0.1
CO_2 from air capture	−0.4	0.3
H_2 from electrolysis + wind	1.0	0.2
H_2 from chlorine electrolysis	1.0	0.2
H_2 from cracking	1.7	1.7
H_2 from electrolysis + photovoltaics	2.4 − 0.6	0.8
H_2 from steam reforming	7.5 − 11.9	1.7
Natural gas	0.5	1.3
Methanol (CH_3OH)	0.7	0.9
Ethylene oxide (C_2H_4O)	1.8	1.4
Propylene oxide (CH_3CHCH_2O)	4.5	2.2

Source: Adapted from Ref. [10].

as compared to the existing or conventional processes, then it should not be considered for further analysis.

3.3. Insights obtained from life cycle assessment results

Interpretation of LCA results from the perspectives of uncertainties owing to missing data or certain assumptions must be properly presented. In this context, once the LCA results are obtained, it is important that we identify the most important parameters and their corresponding values used in our LCA study. These values should always be cross-checked with the available literature as well as the databases. It is also good practice to conduct an uncertainty analysis of the LCA results with respect to these sensitive parameters [23]. The uncertainty analysis can be carried out by setting up Monte Carlo simulations and by setting variations in these parameters [24, 25]. For each of these parameters, we first fix the lower and upper bounds using the data obtained from the literature and databases, then we assume a distribution and generate data samples within these bounds. Once we have data samples for each of the parameters, then we perform LCA studies for all combinations and this results in a ranking of the parameters based on their impact on the LCA results. Besides the sensitivity analysis, different feedstock production technologies should also be considered to assess the influence of potential technology changes and all LCA-related assumptions such as allocation criteria must also be analyzed.

3.4. *Unconventional aspects in the assessment of environmental impacts*

Apart from various types of existing environmental impact models, unconventional metrics can also be accounted for. Such environmental metrics or indexes include time-dependent GHG emissions, water footprint, and several indirect effects like economic or social behavior pertaining to CCU. All these factors taken together along with the other factors discussed previously present the total sustainability metric for any process.

3.4.1. *Time-dependent global warming impacts*

The impact of GHG emissions is also a time-dependent phenomenon. If GHG emissions can be delayed, then in a way they will start absorbing radiations later than they would have if emitted immediately [20]. This concept is essentially important from the perspective of CCU. In CCU, CO_2 emission is delayed by incorporating CO_2 into a value-added product, which will only release this CO_2 once the product is completely used. In the conventional method for evaluating global warming impacts, we assume the radiation is absorbed by the generated GHG emissions over a fixed time horizon, which is usually considered to be 100 years, irrespective of when the GHG is actually emitted. For example, a GHG emitted into the atmosphere today will absorb radiation for the next 100 years. However, if the GHG emission is delayed by say 10 years, it will absorb the same amount of radiation (i.e. if not delayed) in 110 years. Thus, if we assume the time horizon to be 100 years, then in the first 10 years, the GHG does not absorb any radiation as it is not even emitted. Therefore, delaying emissions (as in the case of CCU) seems to be preferable from a global warming mitigation viewpoint. To account for the emission time on evaluating the impact of GHG, revised metrics have been developed [21]. As per heuristic, we assume that the global warming impact of CO_2 emission over a 100-year time horizon is reduced by 1% for every delayed year [10].

3.4.2. Indirect effects in life cycle assessment

The role of an LCA study is to assess the environmental impact. However, we must also understand that the impact on the environment can also be through several indirect means like economic and social factors. For example, the success of CCU is largely governed by the market demand for CO_2-derived products [3, 26]. Dutta *et. al.* [3] showed how replacing a conventional product with a CO_2-derived product can enhance the potential of CO_2 utilization as an option to curb CO_2 emissions. The authors have shown how replacing fossil-based fuels like gasoline and diesel with CO_2-derived products like methanol and dimethyl ether, respectively, can increase the market opportunity for CCU products, which in turn can aid to reduce environmental impacts. Jones *et al.* [27] presented key perspectives on the social acceptance (i.e. socio-political, market, and community acceptance) pertaining to CO_2 utilization. The authors present the significance of social acceptance as an important aspect of the research, development, and deployment of the CO_2 utilization process. The approach to account for all consequences inclusive of the market as well as social factors is known as consequential LCA [28]. However, this concept is still at its nascent stage and further research needs to be carried out in this domain.

4. Life Cycle Assessment for Carbon Capture Utilization Systems: Sample Results

In this section, some sample results of LCA studies pertaining to CCU systems are presented. As CO_2 capture is an integral part of CCU, LCA results for different CO_2 capture options are given in Section 4.1. Following the LCA results for capture processes, an LCA study comparing the environmental impacts of CO_2-based polyol production with the conventional process for polyol production is presented in Section 4.2. Comparative LCA results of various CO_2 utilization options (as stated in Section 2.2) are shown in Section 4.3.

4.1. Life cycle assessment of CO_2 capture processes

Khoo and Tan [12] performed a comparative analysis of the different post-conversion CO_2 capture options. Their results indicated that GWP of CO_2 capture via chemical absorption using monoethanolamine (MEA) solvent was lower than that via cryogenics, membrane separation, or pressure swing adsorption process. Although the chemical absorption process results in the emission of acid gasses, its energy demand is significantly lower than the other options [12]. Figure 2 illustrates the GWP of the different CO_2 capture options like post-conversion, pre-conversion, and oxy-combustion. Each of these options is further classified for pulverized coal (PC), combined cycle gas turbine (CCGT), and integrated coal gasification combined cycle (IGCC) plants and compared against the no CO_2 capture scenario.

The GWP for PC power plants without CO_2 capture is 876 kg CO_2-eq/MWh, while the post-conversion capture via MEA results in about 203 kg CO_2-eq/MWh, and for oxy-fuel combustion, it is only 154 kg CO_2-eq/MWh. In the case of CCGT power plants, the GWP for oxy-fuel combustion is 120 kg CO_2-eq/MWh and 173 kg CO_2-eq/MWh for post-combustion with respect to 471 kg CO_2-eq/MWh without CO_2 capture. The GWP for pre-combustion capture and oxy-fuel combustion in IGCC plants is about 190 kg CO_2-eq/MWh and 200 kg CO_2-eq/MWh, respectively, compared to 1009 kg CO_2-eq./MWh without CO_2 capture. Thus, from the above results [15], it is evident that about 82% reduction in GWP

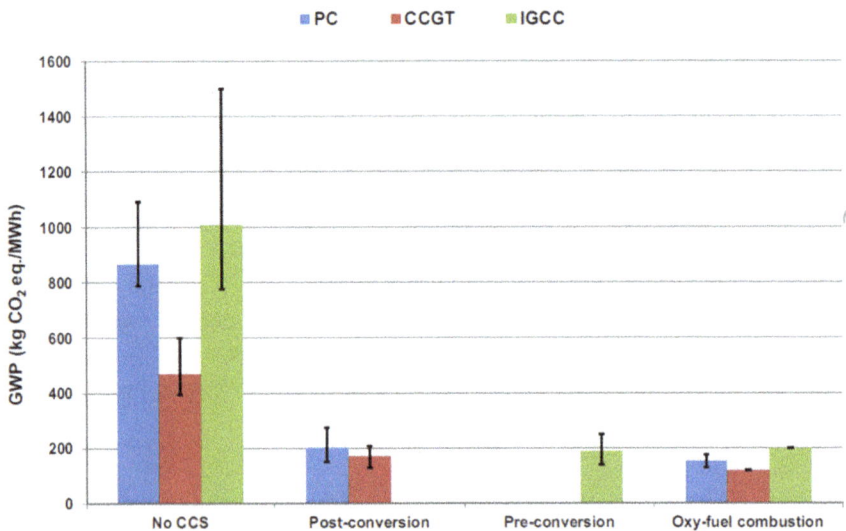

Figure 2.   GWP of different CO_2 capture options for PC, CCGT, and IGCC plants.

Source: Adapted from Ref. [15] with permission from Elsevier.

can be obtained via oxy-fuel combustion in PC and IGCC plants and about 63% via post-combustion capture in CCGT plants.

4.2. *Life cycle assessment of CO_2-based polyol production process*

Polyethercarbonate polyols can be produced by the polymerization of CO_2 and propylene oxide using a double metal cyanide catalyst and multifunctional alcohol (e.g. glycerol) as starters [29]. Thus, polyethercarbonate polyols can be an alternative to the conventional polyether polyols, which had an estimated worldwide production of 8 Mt/a in 2012 [30]. Theoretically, utilizing CO_2 for the production of polyethercarbonate polyols has the potential to utilize about 1.6 Mt/a of CO_2 as feedstock. For the polyethercarbonate polyol production process, CO_2 can be inserted as a monomeric C1 building block, and as a result, the energy-intensive cleavage of C=O bonds can be avoided.

Assen and Bardow [30] performed a cradle-to-gate LCA study for CO_2-based polyol production. The feedstock CO_2 is obtained from a lignite power plant with CO_2 capture. Owing to the CO_2 capture unit, the net electricity output of the lignite power plant is reduced and must be compensated [31]. Thus, the product system for CO_2-based polyols consists of CO_2 sources including CO_2 capture, electricity compensation, and CO_2 utilization for polyol production. This entire product system was referred to as the CCU system. The power plant produced 950 MW of power with an efficiency of 43%. Using an amine-based absorption process for post-combustion CO_2 capture, 300 kg/h of CO_2 was captured from the flue gas [32]. The regeneration heat demand of the capture process was about 2760 MJ per ton CO_2 captured, which corresponds to an equivalent electrical work loss of 123 kWh per ton CO_2 captured [32]. The captured CO_2 was then compressed to 17.5 bar, which consumed an additional 142 kWh of electricity per ton of CO_2. The CO_2 cylinder rags were transported to the 40 km distant polyol production plant by truck. Due to the energy requirement for CO_2 compression and the solvent regeneration process, the electricity output to the grid was reduced by 265 kWh per ton of CO_2 captured.

In an alternative approach reported by NETL [33], the compensated electricity of 0.60 kg CO_2-eq and 0.16 kg oil-eq per kWh was considered. In the production of CO_2-based polyethercarbonate polyols, cyclic propylene carbonate was obtained as a by-product. The amount of by-product was found to vary in the range of 0.02–0.07 kg cyclic propylene carbonate per kg polyol, depending on the CO_2 content in the final polyethercarbonate polyols. The by-product was separated from the polyols by passing the reactor output through a thin-film evaporator. The environmental impact of the CO_2-based polyol production was compared with respect to the production of polyols via the conventional route. Conventionally, polyols are produced using ethylene oxide and propylene oxide in presence of the same double metal cyanide catalyst as in the CCU system [30]. The conventional system included electricity production via a lignite power plant with a net efficiency of 43% but without CO_2 capture resulting in about 0.96 kg CO_2-eq and 0.26 kg oil-eq per kWh [30]. The global warming and fossil resource depletion impacts were evaluated on the basis of 1 kg of polyol production and 0.36 kWh of electricity supplied to the grid.

Figure 3 illustrates the global warming impacts for conventional polyether polyols and CCU systems with polyethercarbonate polyols (20 wt% CO_2). It was observed that in both cases, the production of epoxides

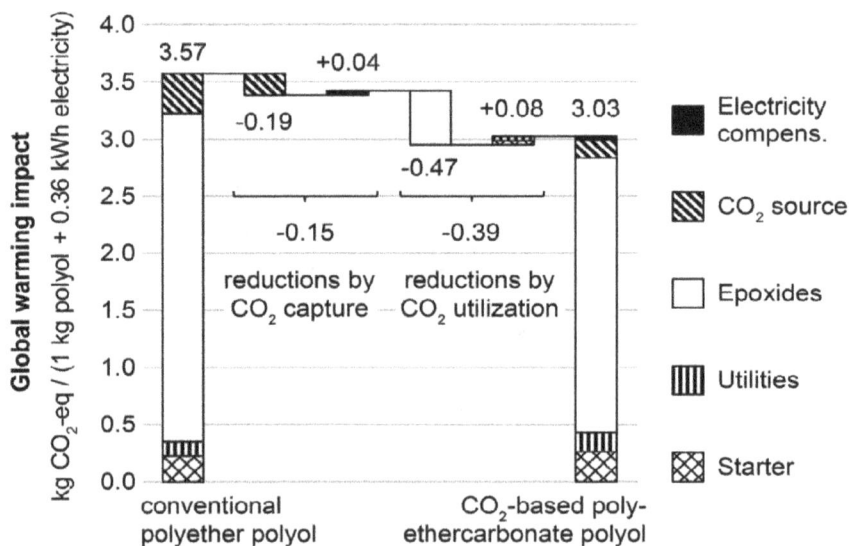

Figure 3.   Global warming impacts in kg CO_2-equivalents of conventional polyether polyols and CO_2-based polyethercarbonate polyols (20 wt% CO_2 content).

Source: Adapted from Ref. [30] with permission from The Royal Society of Chemistry.

contributes toward maximum global warming impact. The CCU system resulted in a 15% reduction, i.e. −0.54 $kgCO_2$-eq/(1 kg polyol + 0.36 kWh electricity) in GHG emissions. For the CCU system, the substitution of emission-intensive epoxides with CO_2 contributed about 72%, i.e. −0.39 $kgCO_2$-eq/(1 kg polyol + 0.36 kWh electricity) toward the reduced GHG emissions and the remaining 28% reduction, i.e. −0.15 $kgCO_2$-eq/ (1 kg polyol + 0.36 kWh electricity) was due to the reduction in emission via CO_2 source and extra emissions for electricity compensation.

Figure 4 illustrates the fossil resource depletion for the conventional polyether polyols and the CCU system with polyethercarbonate polyols containing 10, 20, and 30 wt% CO_2. It was observed that the fossil resource depletion for the CCU system was found to be lower than the conventional process. Substitution of epoxides by CO_2 feedstock in the CCU system was found to be the major contributor toward the reduction in fossil resource depletion. Based on the LCA studies for carbon capture and sequestration system [34], it is observed that the fossil resource demand of power plants increases due to the additional energy require- ment of the CO_2 capture process. However, the increased fossil resource

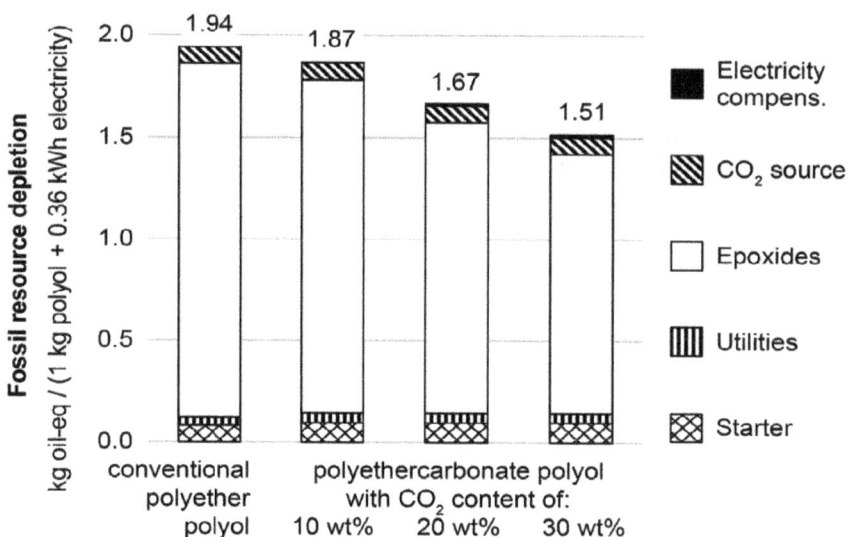

Figure 4. Fossil resource depletion in kg oil-equivalents of conventional polyether polyols and CO_2-based polyethercarbonate polyols with 10, 20, and 30 wt% CO_2 content.
Source: Adapted from Ref. [30] with permission from The Royal Society of Chemistry.

demand from CO_2 capture was compensated by the reduction in the fossil resource depletion owing to the substitution of epoxides by the CO_2 feedstock. As a result, the net reduction in fossil resource depletion is about 22% for a 30 wt% CO_2 content compared to the conventional process [30].

Based on the specific case studies presented, the CO_2-based polyol production process is indeed an excellent option for utilizing CO_2 as a feedstock. It can result in a reduction of up to 3 kg CO_2-eq GHG emissions per kg CO_2 incorporated. CO_2-based polyols can also be readily processed to polyurethanes, thus enabling large amounts of CO_2 utilization and reduction of environmental impacts.

4.3. *Comparative life cycle assessment results for various CO_2 utilization pathways*

The different CO_2 utilization options like EOR, mineralization, and chemical production are compared against CO_2 capture with the single use of 1 ton of CO_2 as the basis for comparison [15]. In this case for CO_2-based chemical production, dimethyl carbonate (DMC) is specifically chosen. Aresta and Galatola [34] showed that DMC production starting from CO_2 resulted in 31 kg CO_2-eq./kg DMC as compared to 132 kg CO_2-eq./kg DMC via the conventional phosgene-based production route. Besides GWP, it was also found that other environmental impacts like acidification potential, ozone depletion potential, eutrophication potential, etc., were also significantly lower for the CO_2-based DMC production as compared to the conventional phosgene-based route. The reduction in environmental impact was mainly due to the significantly lower energy requirement for the production of raw materials used in the CO_2-based synthesis (CO_2 captured via chemical absorption using MEA solvent, ammonia, and hydrogen) compared to the traditional phosgene process (sodium hydroxide and chlorine).

The results of the study carried out by Cuéllar-Franca and Azapagic [15] are compiled in Figure 5. Their work demonstrated that the GWP for CO_2 capture only results in 0.276 kg CO_2-eq/kg CO_2 removed, which comes out to be significantly lower than any CCU options. CO_2-based chemical production (using DMC as the representative) results in about 59.4 kg CO_2-eq/kg CO_2 removed, GWP of biofuels via microalgae ranges from 0.1 to 2.75 kg CO_2-eq/kg CO_2 removed depending on the choice of

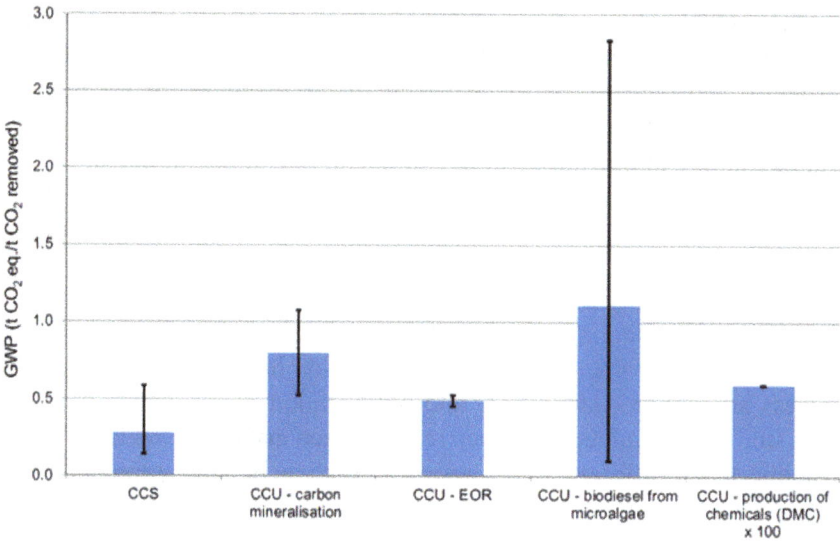

Figure 5.   Comparison of GWP for different CCU options and CO_2 capture only.

Source: Adapted from Ref. [15] with permission from Elsevier.

technology and process operating conditions, GWP of carbon mineralization and EOR is found to be about 0.75 kg CO_2-eq./kg CO_2 removed and 0.5 kg CO_2-eq./kg CO_2 removed, respectively.

5. Conclusion and Future Perspectives

CCU has achieved significant attention in recent times in the context of climate mitigation. The acceptance of the CCU system largely depends on comprehensive LCA studies so that the environmental benefits possessed by any CCU system become evident to policymakers, thus aiding in decision-making. This chapter demonstrated the benefits of LCA investigations for CCU options and how various LCA modeling parameters lead to significant insights in the aim toward a low carbon future. The concept of assessing the total sustainability of a CCU process undoubtedly requires further research in areas such as time-dependent climate conditions, economic evaluation, or social concerns. The inclusion of indirect effects like market potential of a CO_2-derived product must be included within the scope of LCA studies, as the future of any CCU system is strongly governed by the market potential of the product. For example, replacing conventional fuels like gasoline and diesel with CO_2-derived methanol and dimethyl ether will increase the demand for these CO_2-based products, which will undoubtedly enhance the potential of CCU from a global perspective.

Predominantly, research studies in the CCU domain have been focused either on the improvement of catalysts for CCU processes or process simulation and optimization-related studies. However, it is indeed necessary to not only combine the research efforts of catalyst experts with that of process modeling experts but also ensure that the LCA experts are necessarily part of it as well. The amalgamation of catalyst development, process simulation, integration, and optimization along with a detailed LCA study is undoubtedly an open research area that can provide valuable insights about different CCU systems. Such a holistic approach will indeed give a boost to CCU as a promising concept to curb CO_2 emissions and will open up various future research possibilities within the CCU community.

References

[1] M. Aresta, A. Dibenedetto, and A. Angelini, Catalysis for the valorization of exhaust carbon: From CO_2 to chemicals, materials, and fuels. technological use of CO_2. *Chem. Rev.* **114**, 1709–42 (2014). doi: 10.1021/cr4002758.

[2] K. Armstrong and P. Styring, Assessing the potential of utilization and storage strategies for post-combustion CO_2 emissions reduction. *Front. Energ. Res.* **3**, 1–9 (2015). doi: 10.3389/fenrg.2015.00008.

[3] A. Dutta, S. Farooq, I. A. Karimi, and S. A. Khan, Assessing the potential of CO_2 utilization with an integrated framework for producing power and chemicals. *J. CO_2 Util.* **19**, 49–57 (2017). doi: 10.1016/j.jcou.2017.03.005.

[4] K. Roh, A. S. Al-Hunaidy, H. Imran, and J. H. Lee, Optimization-based identification of CO_2 capture and utilization processing paths for life cycle greenhouse gas reduction and economic benefits. *AIChE J.* **65**, 1–15 (2019). doi: 10.1002/aic.16580.

[5] S. Vasudevan, S. Farooq, I. A. Karimi, M. Saeys, M. C. G. Quah, and R. Agrawal, Energy penalty estimates for CO_2 capture: Comparison between fuel types and capture-combustion modes. *Energy* **103**, 709–714 (2016). doi: 10.1016/j.energy.2016.02.154.

[6] M. Aresta, A. Dibenedetto, and A. Angelini, The changing paradigm in CO_2 utilization. *J. CO_2 Util.* **3–4**, 65–73 (2013). doi: 10.1016/j.jcou.2013.08.001.

[7] M. Aresta, A. Dibenedetto, and A. Dutta, Energy issues in the utilization of CO_2 in the synthesis of chemicals: The case of the direct carboxylation of alcohols to dialkyl-carbonates. *Catal. Today.* **281**, 345–351 (2016). doi: 10.1016/j.cattod.2016.02.046.

[8] C. Jiménez-González, D. J. C. Constable, and C. S. Ponder, Evaluating the "Greenness" of chemical processes and products in the pharmaceutical industry — a green metrics primer. *Chem. Soc. Rev.* **41**, 1485–1498 (2012). doi: 10.1039/c1cs15215g.

[9] D. J. C. Constable, A. D. Curzons, and V. L. Cunningham, metrics to "green" chemistry — Which are the best? *Green Chem.* **4**, 521–527 (2002). doi: 10.1039/b206169b.

[10] N. Von Der Assen, P. Voll, M. Peters, and A. Bardow, Life cycle assessment of CO_2 capture and utilization: A tutorial review. *Chem. Soc. Rev.* **43**, 7982–7994 (2014). doi: 10.1039/c3cs60373c.

[11] L. J. Müller, A. Kätelhön, M. Bachmann, A. Zimmermann, A. Sternberg, and A. Bardow, A guideline for life cycle assessment of carbon capture and utilization. *Front. Energy Res.* **8**, 1–20 (2020). doi: 10.3389/fenrg.2020.00015.

[12] H. H. Khoo and R. B. H. Tan, Life cycle investigation of CO_2 recovery and sequestration. *Environ. Sci. Technol.* **40**, 4016–4024 (2006). doi: 10.1021/es051882a.

[13] G. Garcia-Garcia, M. C. Fernandez, K. Armstrong, S. Woolass, and P. Styring, Analytical review of life-cycle environmental impacts of carbon capture and utilization technologies. *Chem. Sus. Chem.* **14**, 995–1015 (2021). doi: 10.1002/cssc.202002126.

[14] H. H. Khoo, J. Bu, R. L. Wong, S. Y. Kuan, and P. N. Sharratt, Carbon capture and utilization: Preliminary life cycle CO_2, energy, and cost results of potential mineral carbonation. *Energy Procedia* **4**, 2494–2501 (2011). doi: 10.1016/j.egypro.2011.02.145.

[15] R. M. Cuéllar-Franca and A. Azapagic, Carbon capture, storage and utilisation technologies: A critical analysis and comparison of their life cycle environmental impacts. *J. CO_2 Util.* **9**, 82–102 (2015). doi: 10.1016/j.jcou.2014.12.001.

[16] B. Singh, A. H. Strømman, and E. G. Hertwich, Comparative life cycle environmental assessment of CCS technologies. *Int. J. Greenh. Gas. Control* **5**, 911–921 (2011). doi: 10.1016/j.ijggc.2011.03.012.

[17] S. Krishnamurthy, V. R. Rao, S. Guntuka, P. Sharratt, R. Haghpanah, A. Rajendran, M. Amanullah, I. A. Karimi, and S Farooq, CO_2 capture from dry flue gas by vacuum swing adsorption: A pilot plant study. *AIChE J.* **60**, 1830–1842 (2014). doi: 10.1002/aic.14435.

[18] P. Markewitz, W. Kuckshinrichs, W. Leitner, J. Linssen, P. Zapp, R. Bongartz, A. Schreibera, and T. E. Müller, Worldwide innovations in the

development of carbon capture technologies and the utilization of CO_2. *Energy Environ. Sci.* **5**, 7281–7305 (2012). doi: 10.1039/c2ee03403d.

[19] M. Pehnt and J. Henkel, Life cycle assessment of carbon dioxide capture and storage from lignite power plants. *Int. J. Greenh. Gas Control* **3**, 49–66 (2009). doi: 10.1016/j.ijggc.2008.07.001.

[20] N. Von Der Assen, J. Jung, and A. Bardow, Life-cycle assessment of carbon dioxide capture and utilization: Avoiding the pitfalls. *Energy Environ. Sci.* **6**, 2721–2734 (2013). doi: 10.1039/c3ee41151f.

[21] G. P. Peters, B. Aamaas, T. M. Lund, C. Solli, and J. S. Fuglestvedt, Alternative "global warming" metrics in life cycle assessment: A case study with existing transportation data. *Environ. Sci. Technol.* **45**, 8633–8641 (2011). doi: 10.1021/es200627s.

[22] Ecoinvent (n.d.) (Accessed June 3, 2021), https://www.ecoinvent.org/.

[23] A. E. Björklund, Survey of approaches to improve reliability in LCA. *Int. J. Life Cycle Assess.* **7**, 64–72 (2002). doi: 10.1007/BF02978849.

[24] A. Dutta, I. A. Karimi, and S. Farooq, Technoeconomic perspective on natural gas liquids and methanol as potential feedstocks for producing olefins. *Ind. Eng. Chem. Res.* **58**, 963–972 (2019). doi: 10.1021/acs. iecr.8b05277.

[25] J. Jung, N. Von Der Assen, and A. Bardow, Sensitivity coefficient-based uncertainty analysis for multi-functionality in LCA. *Int. J. Life Cycle Assess.* **19**, 661–676 (2014). doi: 10.1007/s11367-013-0655-4.

[26] E. Masanet, Y. Chang, A. R. Gopal, P. Larsen, W. R. Morrow, R. Sathre, A. Shehabi, and P. Zhai, Life-cycle assessment of electric power systems. *Annu. Rev. Environ. Resour.* **38**, 107–136 (2013). doi: 10.1146/annurev-environ-010710-100408.

[27] C. R. Jones, B. Olfe-Kräutlein, H. Naims, and K. Armstrong, The social acceptance of carbon dioxide utilisation: A review and research Agenda. *Front. Energy Res.* **5**, 1–13 (2017). doi: 10.3389/fenrg.2017.00011.

[28] J. M. Earles and A. Halog, Consequential life cycle assessment: A review. *Int. J. Life Cycle Assess.* **16**, 445–453 (2011). doi: 10.1007/s11367-011-0275-9.

[29] J. Langanke, A. Wolf, J. Hofmann, K. Böhm, M. A. Subhani, T. E. Müller, W. Leitner, and C. Gürtler, Carbon dioxide (CO_2) as sustainable feedstock for polyurethane production. *Green Chem.* **16**, 1865–1870 (2014). doi: 10.1039/c3gc41788c.

[30] N. Von Der Assen and A. Bardow, Life cycle assessment of polyols for polyurethane production using CO_2 as feedstock: Insights from an industrial case study. *Green Chem.* **16**, 3272–3280 (2014). doi: 10.1039/c4gc00513a.

[31] R. Sathre, M. Chester, J. Cain, and E. Masanet, A framework for environmental assessment of CO_2 capture and storage systems. *Energy* **37**, 540–548 (2012). doi: 10.1016/j.energy.2011.10.050.

[32] P. Moser, S. Schmidt, G. Sieder, H. Garcia, T. Stoffregen, and V. Stamatov, The post-combustion capture pilot plant Niederaussem — Results of the first half of the testing programme. *Energy Procedia* **4**, 1310–1316 (2011). doi: 10.1016/j.egypro.2011.01.188.

[33] National Energy Technology Laboratory LCA, Plant EPC (EXPC) P. DOE/NETL- 403-110809. (Accessed June 3, 2021), http://www.netl.doe.gov/ea/about.

[34] M. Aresta and M. Galatola, Life cycle analysis applied to the assessment of the environmental impact of alternative synthetic processes. The dimethyl-carbonate case: Part 1. *J. Clean. Prod.* **7**, 181–193 (1999). doi: 10.1016/S0959-6526(98)00074-2.

Chapter 5

Green Principles in Active Pharmaceutical Ingredient Manufacturing as Seen through the Lens of Life Cycle Assessment

Valerio ISONI[*,‡], **Hsien H. KHOO**[*,§] **and Alvin W. L. EE**[†,¶]

Institute of Chemical and Engineering Sciences (ICES), Agency for Science, Technology and Research (A*STAR), Singapore 627833

†*Energy Studies Institute, National University of Singapore (NUS), 29 Heng Mui Keng Terrace, #0-01, Singapore 119620*

‡*isoniva@ices.a-star.edu.sg*

§*khoo_hsien_hui@ices.a-star.edu.sg*

¶*esiaewl@nus.edu.sg*

Since its conception over a decade ago, *green chemistry* concepts emulate environmentally benign design principles such as non-hazardous chemical reactions, alternative synthetic pathways for pollution prevention, or analytical areas of chemistry research aimed at improving organic synthesis. Numerous applications have evolved where green chemistry has spread beyond research laboratories to industrial applications. However, efforts are required, particularly in the area of LCA, to evaluate the environmental impact of "green" applications through the lens of a holistic perspective. This chapter looks into five selected green chemistry principles to evaluate the "green credentials" of the manufacturing of an Active Pharmaceutical Ingredient (API).

1. Introduction

At the onset of the 21st century, the concept of "environmental sustainability" has become widely recognized and increasingly accepted by various stakeholders as a guiding principle for both decision-making and corporate strategies. Along with the emergence of environmental sustainability, green chemistry has become a popular theme for the purpose of carrying out chemical activities without causing harm to the environment or without the use and generation of hazardous substances. Since its conception over a decade ago, *green chemistry* concepts emulate environmentally benign design principles such as non-hazardous chemical reactions, alternative synthetic pathways for pollution prevention, or analytical areas of chemistry research aimed at improving organic synthesis [1, 2]. The 12 principles of green chemistry are summarized in Table 1.

Born as guidelines to push for more environmentally benign, safer, and more efficient chemical processes in both academia and industry, the 12 principles managed to influence the way chemists perform their

Table 1.   Compilation of the 12 green chemistry principles.

1	Waste Prevention
2	***Atom Economy***
3	***Less Hazardous Chemical Synthesis***
4	Designing Benign (Safer) Chemicals
5	***Safer Solvents and Auxiliaries***
6	Design for Energy Efficiency
7	***Use of Renewable Feedstocks***
8	Reduce Derivatives
9	Catalysis
10	Design for Degradation
11	Real Time analysis for Pollution Prevention
12	***Inherently Safer Chemistry for Accident Prevention***

Note: The principles used for the case study are highlighted in bold italic reactions [3, 4]. With greater awareness, more questions started to arise within the research community, including "Which principles are the most important when evaluating the greenness of a chemical process? and "Can green principles lead to environmental pitfalls if followed in isolation?".

One of the underlying challenges found in the application of green chemistry in the high-value chemical and pharmaceutical industry is in determining the relative importance and impact of each principle. In this work, we provide a different perspective of some selected green chemistry applications which are traditionally targeted at the molecular or chemical synthesis level and focus on a broader system-wide aspect. The authors selected four green chemistry principles to evaluate the "green credentials" of the manufacturing of an Active Pharmaceutical Ingredient (API) via a life cycle assessment (LCA) approach.

1.1. *Case study example: clopidogrel's API manufacture*

The production of clopidogrel's API was selected as a case study to test the outcome of green chemistry under the lenses of four different green chemistry principles. In the manufacturing of this API, the main chemicals and solvents by mass contribution required to produce 1 kg of clopidogrel are listed in Table 2. A simplified system boundary is illustrated in Figure 1.

A set of life cycle environmental impacts will be tested with the following green chemistry principles applied within the LCA model of API manufacture.

Table 2.   List of main material input per
1 kg of clopidogrel's API.

Material/Chemical	Mass (kg)
Methanol	33.0
Dichloromethane (DCM)	45.8
Acetone	31.0
Di-isopropyl ether	5.8
Iso-propanol	3.0
Dimethyl sulfate (DMS)	0.9
Dimethyl sulfoxide	0.8
Sodium cyanide	0.40

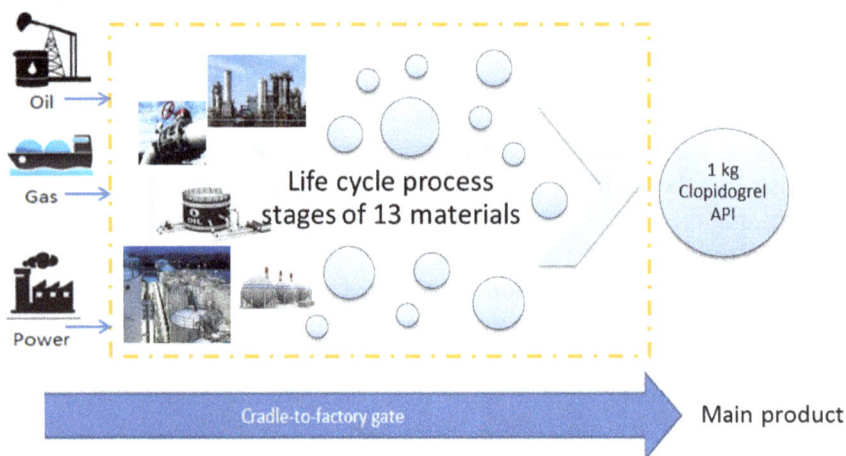

Figure 1. Main model: simplified LCA system boundary of conventional API manufacture.

2. Green Chemistry Principle Applications

Overall, green chemistry seeks to reduce chemical hazards and increase the environmental sustainability performance of manufacturing processes via adopting environmentally benign practices [5, 6]. However, following individual guiding principles in isolation could unexpectedly lead to less than ideal outcomes in terms of safety, sustainability, and efficiency.

Based on LCA concepts, the four selected green chemistry principles tested for our case study (highlighted in bold in Table 1) are atom economy (P2), less hazardous synthesis/accident prevention (P3/P12), designing benign solvents (P5), and use of renewable feedstocks (P7). This selection of principles covers aspects such as manufacturing efficiency and profit (P2), safety (P3, P5), waste generation (P2, P3), and feedstock sustainability (P7).

2.1. *Principle 2: Atom economy evaluation of dimethyl sulfate production*

The intention of green chemistry principle 2 (P2) is to optimize reaction and processing conditions to maximize the use of raw materials (reactants) in the final product, reducing the amount of waste generated. Synthesis of APIs is usually complex involving several reactions,

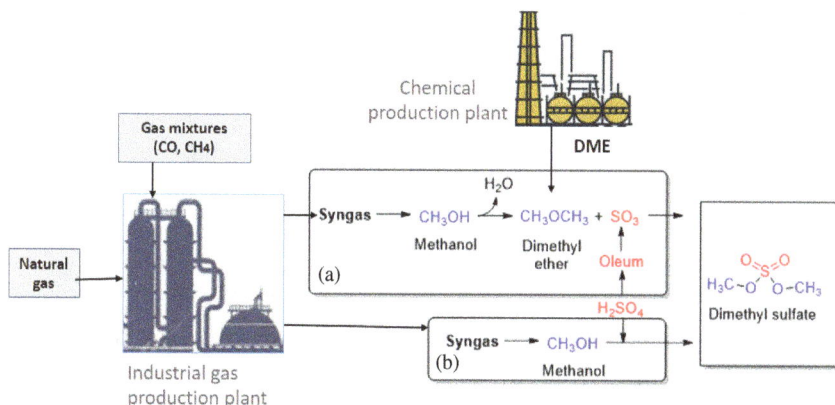

Figure 2.   Production of (a) DMS from dimethyl ether and SO_3 (industrial production route); (b) DMS from methanol and H_2SO_4.

separations, and purification steps. Each of these operations requires solvents and reagents that can be produced or manufactured in more than a single way. The simpler the molecular structure (e.g. solvents), the more variation is possible. This poses an additional challenge to use LCA for comparisons, since multiple ramifications can lead to different results. As an example in this section, atom economy is reviewed from a life cycle perspective for DMS which is one of the chemicals reported in Table 1.

Typically, DMS is manufactured from the reaction of dimethyl ether with sulfur trioxide. As a prior synthesis step, dimethyl ether is produced from the dehydration of methanol, which is industrially produced from syngas, and to complete the life cycle production chain, the industrial production of sulfur trioxide happens from oleum and sulfuric acid. Alternatively, DMS can also be produced directly via the esterification of sulfuric acid with two moles of methanol (Figure 2).

2.2. Principles 3/12: less hazardous synthesis of sodium cyanide/accident prevention by use of safer chemicals

In green chemistry principle 3 (P3), synthetic methods should be designed to use and generate substances that possess little or no toxicity to human health and the environment. In a lab-scale experimental study, Attah-Daniel *et al.* [7] reported the conversion of cassava waste into sodium

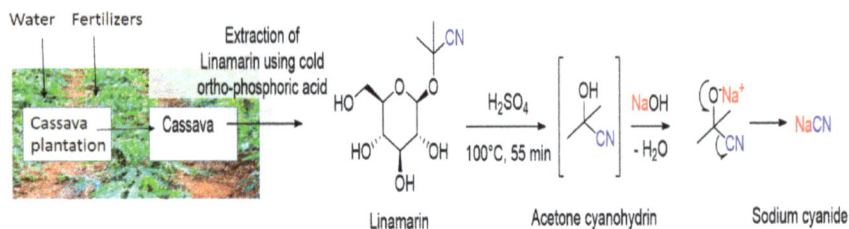

Figure 3.   Sodium cyanide production from cassava leaves.

cyanide from concentrated cyanogenic glucosides present in the plant. Cassava is known to contain cyanogenic glycosides, mainly linamarin (*ca.* 90% total cyanogen), especially the leaves [8–9]. The authors claimed that this process can possibly bypass the need for any prior stage for the production of hydrogen cyanide (HCN), which is typically manufactured from methane and ammonia. This also relates to principle 12 (prevention of accidents) since any potential occupational hazards or accidents that may occur during the industrial production of HCN can be avoided [10, 11].

While the claims for safer production of sodium cyanide under the provided laboratory protocol need further study, it provides an interesting alternative to established manufacturing processes that is also based on renewable feedstock. Based on the cassava leaves characteristics, the authors managed to achieve conversion of linamarin to acetone cyanohydrin and yield sodium cyanide by treatment with sodium hydroxide. The main life cycle stages and chemical transformations are shown in Figure 3.

2.3. *Principle 5: Use of benign solvent*

Alfonsi *et al.* [12] recommended a list of solvent replacement guidelines according to the following assessment criteria: (i) work-related safety; (ii) process safety (including flammability and potential for high emissions through high vapor pressure), and (iii) environmental and regulatory considerations (including potential EHS regulatory restrictions, ozone depletion potential, and photo-reactive potential). The recommended solvent selected for replacing dichloromethane is ethyl acetate and for the replacement of di-isopropyl ether, 2-dimethyl tetrahydrofuran (2-MeTHF). It is worth noting the 2-MeTHF is a solvent which is produced from

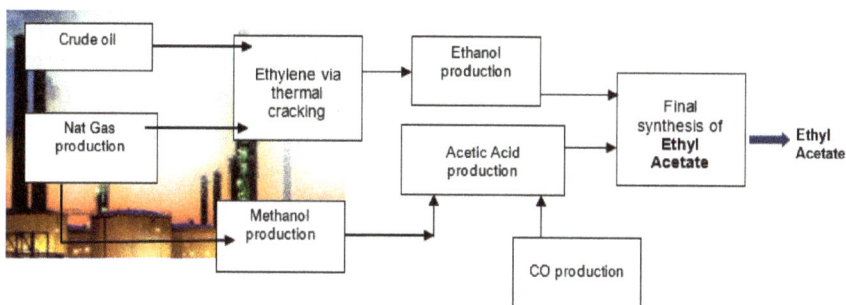

Figure 4. Cradle-to-gate production of ethyl acetate.

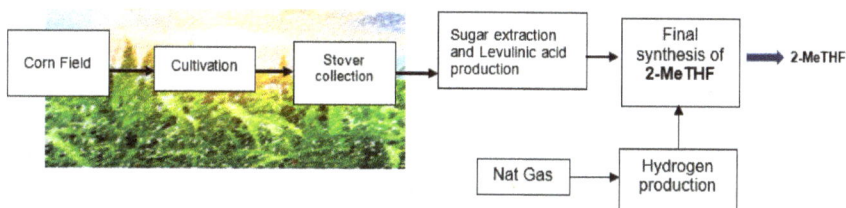

Figure 5.   Cradle-to-gate production of 2-MeTHF.

biomass resources (bio-based). To date, however, the actual definition of a green solvent is still a matter of debate [13, 14]. Here, the LCA comparisons were made between DCM and ethyl acetate as well as di-isopropyl ether and 2-MeTHF to test the environmental impacts of principle 5. The cradle-to-gate LCA models of ethyl acetate and 2-MeTHF are illustrated in Figures 4 and 5, respectively.

2.4. *Principle 7: Use of renewable feedstocks*

The utilization of biomass feedstock for producing fuels and chemicals forms the underlying precept of P7 [15]. For this green chemistry principle investigation, we evaluated the impact of replacing fossil-based methanol and acetone with their bio-derived versions. Several methods to efficiently convert lignocellulosic biomass to usable C_5 and C_6 sugars have been extensively studied for the purpose of producing chemicals, fuels, and materials in a sustainable manner. Among the bio-processes that involve the conversion of fermentable sugars into higher-value chemicals, the A–B–E (acetone–butanol–ethanol) process is introduced as one of the biotechnologies employed for producing bio-acetone. In an example described by Moradi *et al.* [16], 1 kg of straw was successfully converted into 44 g of butanol and 17 g of acetone using *Clostridium acetobutylicum* for the enzymatic transformation. Details of bagasse-to-methanol conversion via gasification were reported by Renó *et al.* [17] as well as Hamelinck and Faaij [18].

3. Results and Discussion

The following environmental impacts were generated for the above-mentioned green chemistry scenarios: Global Warming Potential (GWP), Acidification Potential (AP), Eutrophication Potential (EP), Human Toxicity Potential (HTP), Photochemical Ozone Creation Potential (POCP), Energy Use, and Water Use. The LCIA results are compiled in Figure 6.

The findings of the LCA impacts are interpreted to recognize the benefits as well as downsides of green chemistry applications. In P2, while atom economy is used as a measure of reaction efficiency, a complete picture of the life cycle environmental impacts is not within its scope. From Figure 6(a), the environmental benefits of DMS produced via an alternative production route (esterification of sulfuric acid with two moles of methanol) are obvious. The significant environmental reductions are 95% eutrophication, 83% human toxicity, and 69% GWP. In the test of P3/P12, POCP and human toxicity are significantly reduced in the production of sodium cyanide (NaCN) from cassava waste [7], as compared to NaCN being manufactured from HCN. In terms of safety and hazard prevention [10, 11], this option is preferred. However, an increase of around 54% and over 80% for AP and EP impacts is observed, respectively, as shown in Figure 6(b); this is likely due to the effects of emissions from cassava farming activities involving N and P fertilizers [19, 20].

Figures 6(c) and (d) both demonstrate the quantified impact results for the replacement of DCM with ethyl acetate and di-isopropyl ether with 2-MeTHF. As seen in Figure 6(c), the overall environmental performance of ethyl acetate seems to be more favorable as compared to the use of DCM. On the other hand, during the production of stover to 2-MeTHF, levulinic acid was used as the intermediate step between stover and the final bio-based solvent. The additional steps involve farming and conversion of lignocellulose to usable sugars, which lead to these impacts. More details can be found in Khoo *et al.* [20].

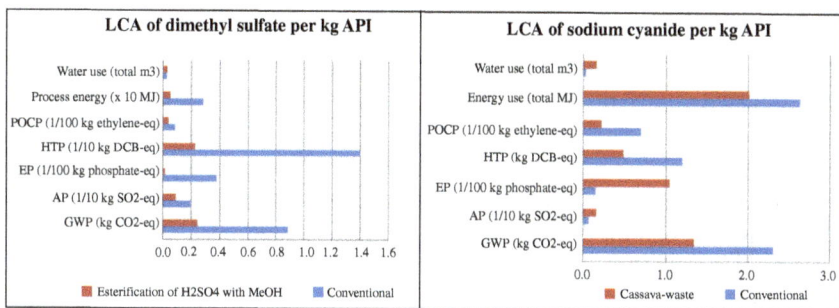

(a) P2: Alternative production route of DMS

(b) P3/P12: Production of sodium cyanide from cassava waste

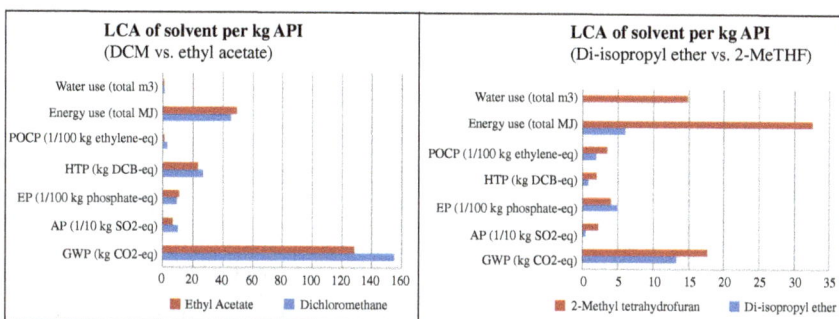

(c) P5: Replacement DCM with ethyl acetate

(d) P5: Replacement of di-isopropyl ether with 2-MeTHF

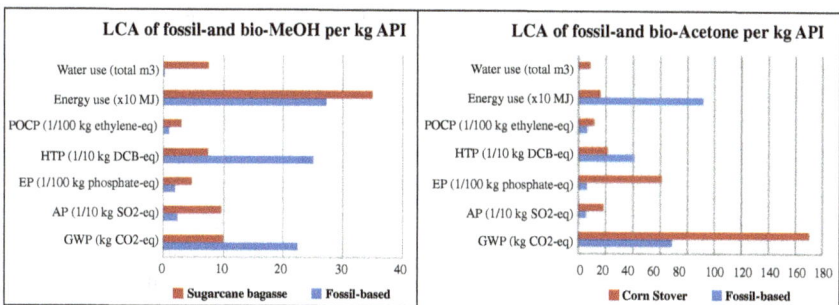

(e) P7: Fossil-based methanol & bio-methanol

(f) P7: Fossil-based acetone & bio-acetone from wheat straw

Figure 6. (a)–(f) LCIA comparisons of selected green chemistry principles per kg API production.

The rest of the results show how holistic systems thinking — via the LCA approach — could help the prevention of unintended or unforeseen consequences in the sustainability assessment of green chemistry [6, 21, 22].

Measures to determine the "greenness" incorporated in any processes or products have been a topic of debate and discussions, where the selection of appropriate methods to measure the environmental performance of a product or process can result in different implications or meanings of sustainability. However, the terms for sustainability or green processes are still vaguely defined. A qualitative summary of the results is shown in Table 3.

Table 3.   Summary of relative comparison of green chemistry applications.

Green Chemistry Principle	Changes in Environmental Impact Categories						Brief Description and Comments
	GWP	AP/EP	HTP	POCP	Energy	Water	
P2	∇	∇	∇	◊	◊	◊	Apart from benefits displayed by the life cycle atom economy, environmental impacts of the alternate routes decreased due to more efficient resource consumption along the production chain.
P3/12	◊	•	∇	∇	◊	•	LCA model of cassava waste-to-sodium cyanide conversion based on experimental case study [7]. Process information to test/ enable scale-up recommended
P5	∇	∇	∇	∇	•	∇	DCM replaced by ethyl acetate as recommended by [12]; further experimental work is recommended to test the effectiveness of solvent replacement
P5	•	•	•	•	•	•	LCA details of 2-MeTHF can be found in [20].
P7	◊	•	∇	•	•	•	LCA of bio-methanol taken from [17, 18]; biggest impact comes from AP/EP and water use.
P7	•	• •	◊	•	◊	• •	Based on ABE technology reported in [16]

Note: • Increase; ◊ Slightly reduced; ∇ Significant. reduction

4. Concluding Remarks and Suggestions

LCA is recommended to be used to measure the "greenness" of green chemistry by uncovering environmental impacts that might not be considered. Holistic sustainability methods and tools that incorporate different environmental effects can help identify important issues that otherwise can be overlooked. As viewed from a larger perspective, Iles and Mulvihill [21] and other literature studies [22, 23] have illustrated that despite the intentions to create greener processes, green chemistry can still generate adverse consequences. Alternatively, Jiménez-González *et al.* [24] stressed the need to move toward metrics based on LCA methodologies as the standard early assessment of synthetic routes for APIs.

By nature of its methodology and application, LCA does not deal directly with complex chemistry and changes at the molecular level. During the conception of the LCA framework, topics involving complex chemistry synthesis were not typically included within its scope of analysis. Therefore, in order to generate appropriate data compiled for complete LCA modeling in chemical or pharmaceutical industries, it is important that a multi-disciplinary team of chemists and engineers work together [24, 25] (Figure 7).

There are numerous applications where green chemistry has marched beyond the research laboratories and found commercial applications. However, lot more efforts are required, particularly in LCA applications for the evaluation of potential and unintended environmental impacts of the various "green" API production systems. Ratti [26] elaborated that developing a successful green process is not only about green chemistry as a single topic but also expand to involve the knowledge of green

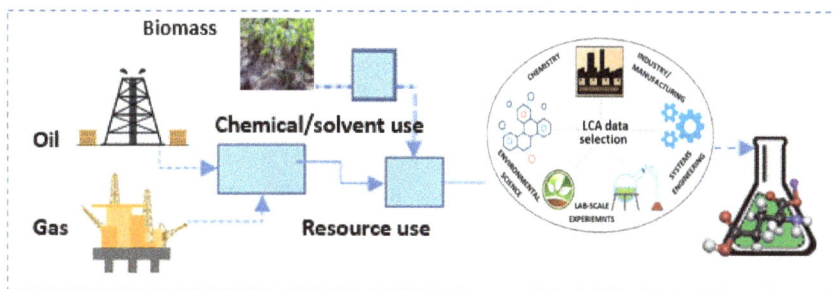

Figure 7.   Multi-disciplinary team required to generate information for LCA modeling involving green chemical synthesis [25].

engineering, biotechnology, economics, and above all toxicology effects [26]. This type of multi-disciplinary activity can be time- and resource-intensive, but it is a necessary effort to accurately comprehend the type of changes in a chemical process that would ultimately lead to an overall environmental gain compared to the *status quo*.

References

[1] P. T. Anastas and R. L. Lankey, Life cycle assessment and green chemistry: The yin and yang of industrial ecology. *Green Chem.* **2**, 289–295 (2000).

[2] P. Tundo, P. Anastas, D. StC. Black, J. Breen, T. J. Collins, S. Memoli *et al.*, Synthetic pathways and processes in green chemistry. *Pure Appl. Chem.* **72**, 1207–1228 (2000).

[3] A. Laurent A, S. I. Olsen, and M. Z. Hauschild, Limitations of carbon footprint as indicator of environmental sustainability. *Environ. Sci. Technol.* **46**, 4100–4108 (2012).

[4] A. Baral, B. R. Bakshi, and R. L. Smith, Assessing resource intensity and renewability of cellulosic ethanol technologies using Eco-LCA. *Environ. Sci. Technol.* **46**, 2436–2444 (2012).

[5] P. J. Dunn, A. S. Wells, and M. T. Williams (Eds.), *Green Chemistry in the Pharmaceutical Industry* (Wiley-VCH Verlag GmBH & Co, Weinheim, 2010).

[6] C. Jiménez-González, D. J. C. Constable, and C. S. Ponder, Evaluating the "Greenness" of chemical processes and products in the pharmaceutical industry — A green metrics primer. *Chem. Soc. Rev.* **41**, 1485–1498 (2012).

[7] B. E. Attah-Daniel, K. Ebisike, C. E. Adeeyinwo, A. R. Adetunji, S. O. O. Olusunle, and O. O. Adewoye, Production of sodium cyanide from cassava wastes. *Int. J. Sci. Technol.* **2**, 707–709 (2013).

[8] N. Phambu, A. S. Meya, E. B. Djantou, E. N. Phambu, P. Kita-Phambu, and L. M. Anovitz, Direct detection of residual cyanide in cassava using spectroscopic techniques. *J. Agric. Food Chem.* **55**, 10135–10140 (2007).

[9] M. R. Haque and J. H. Bradbury, Preparation of linamarin from cassava leaves for use in a cassava cyanide kit. *Food Chem.* **85**, 27–29 (2004).

[10] K. K. Lam and F. L. Lau, An incident of hydrogen cyanide poisoning. *American J. Emergency Med.* **18**(12), 172–175 (2000).

[11] D. Gidlow, Hydrogen cyanide — An update. *Occupational Med.* **67**(90), 662–663 (2017).

[12] K. Alfonsi, J. Colberg, P. J. Dunn, T. Fevig, S. Jennings, T. A. Johnson *et al.*, Green chemistry tools to influence a medicinal chemistry and research chemistry based organization. *Green Chem.* **10**, 31–36 (2008).

[13] C. Capello, U. Fischer, and K. Hungerbuhler, What is a green solvent? A comprehensive framework for the environmental assessment of solvents. *Green Chem.* **9**, 927–934 (2007).

[14] V. Isoni, L. L. Wong, H. H. Khoo, I. Halim, and P. N. Sharratt, Q-SA√ESS: a methodology to help solvent selection for pharmaceutical manufacture at the early process development stage. *Green Chem.* **18**, 6564–6572 (2016).

[15] T. T. H. Nguyen, Y. Kikuchi, M. Noda, and M. Hirao, A new approach for the design and assessment of bio-based chemical processes toward sustainability. *Ind. Eng. Chem. Res.* **54**, 5494–5504 (2015).

[16] F. Moradi, H. Amiri, S. Soleimanian-Zad, M. R. Ehsani, and K. Karimi, Improvement of acetone, butanol and ethanol production from rice straw by acid and alkaline pretreatments. *Fuel* **112**, 8–13 (2013).

[17] M. L. G. Renó, E. E. S. Lora, and J. C. E. Palacio, Life cycle assessment of methanol production from sugarcane bagasse considering two different alternatives of energy supply. *Proc. 20th Intl. Congress of Mech. Eng.*, Nov 15–20, Brazil (2009).

[18] C. N. Hamelinck and A. C. Faaij, Future prospects for the production of methanol and hydrogen from biomass. *J. Power Sources* **111**, 1–22 (2002).

[19] C. P. Mainstone and W. Parr, Phosphorus in rivers-ecology and management. *Sci. Tot. Environ.* **282–283**, 25–47 (2002).

[20] H. H. Khoo, L. L. Wong, J. Tan, V. Isoni, and P. Sharratt, Synthesis of 2-Methyl tetrahydrofuran from various lignocellulosic feedstocks: Sustainability assessment via LCA. *Res. Conserv. Rec.* **95**, 174–182 (2015).

[21] A. Iles and M. J. Mulvihill, Collaboration across disciplines for sustainability: Green chemistry as an emerging multistakeholder community. *Environ. Sci. Technol.* **46**, 5643–5649 (2012).

[22] W. De Soete, S. Debaveye, S. De Meester, G. Van der Vorst, W. Aelterman, B. Heirman *et al.*, Environmental sustainability assessments of pharmaceuticals: An emerging need for simplification in life cycle assessments. *Environ. Sci. Technol.* **48**, 12247–12255 (2014).

[23] D. Tanzil and B. R. Beloff, Assessing impacts: Overview on sustainability indicators and metrics. *Environ. Quality Manage.* **15**, 41–56 (2006).

[24] C. Jiménez-González, C. Ollech, W. Pyrz, D. Hughes, Q. B. Broxterman, and N. Bhathela, Expanding the boundaries: Developing a streamlined tool for eco-footprinting of pharmaceuticals. *Org. Process Res. Dev.* **17**(2), 239–246 (2013).

[25] H. H. Khoo, V. Isoni, and P. N. Sharratt, LCI data selection criteria for a multidisciplinary research team: LCA applied to solvents and chemicals. *Sustain. Prod. Consump.* **16**, 68–87 (2018).

[26] R. Ratti, Industrial applications of green chemistry: Status, challenges and prospects. *SN Appl. Sci.* **2**, 263 (2020).

Chapter 6

Life Cycle-Atom Economy and Life Cycle Assessment as a Hybrid Sustainability Assessment Tool

Pancy ANG[*,‡]**, Poe Ronald Hanniel BUSTAMANTE**[†,§]**,**
Praveen THONIYOT[*,¶] **and Hsien H. KHOO**[*,‖]

*Institute of Chemical and Engineering Sciences (ICES), Agency for Science, Technology and Research (A*STAR), Singapore 627833*

†*Newcastle University, Singapore 599493*

‡*pancy_ang@ices.a-star.edu.sg*

§*r.h.b.poe2@newcastle.ac.uk*

¶*thoniyotp@ices.a-star.edu.sg*

‖*khoo_hsien_hui@ices.a-star.edu.sg*

While Atom Economy (AE) is an important concept in green chemistry, the spectrum it covers is limited as raw materials that make up the reactants of a production process are not accounted for. Life Cycle (LC) concepts derived from Life Cycle Assessment (LCA) are therefore incorporated with AE to form an implicit LC-AE framework for a holistic sustainable assessment of a production route. AE and LC-AE of the production routes for four selected industrial solvents, Tetrahydrofuran (THF), 2-Methyltetrahydrofuran (2-MeTHF), *N,N*-Dimethylformamide (DMF), and Dimethyl sulfoxide (DMSO), are investigated. A decrease in LC-AE percentages from AE percentages is observed which indicates that the processes are more inefficient when considering the production route's cradle. LC-AE of the alternatives for THF and DMF, namely 2-MeTHF and DMSO, is also covered to understand the potential of the latter in replacing the solvents. The alternatives achieved higher efficiency which resulted in a higher LC-AE. The LCA comparison of Production Route 5 (DMF production) and Production Route 7 (DMSO production) further affirmed that DMSO is more environmentally friendly than DMF. Environmental impact reductions of all the selected environmental impact categories are achieved when production of 1 kg of DMSO is used as a substitute for 1 kg of DMF.

1. Introduction

Atom Economy (AE) was first introduced by Barry Trost in 1991 [1] to indicate the efficiency of a process in converting the number of atoms of participating reactants into the desired product. This simple percentage metric can be determined easily from reaction equations, which provided a way to assess whether the transformation follows green chemistry principles. As per green chemistry principles, highly efficient (high AE) processes are preferred for sustainability reasons [2]. Even though AE provides a good matrix for understanding the efficiency of a given set of chemical transformations, considering AE alone may not be a good representation toward sustainability since the precedent processes are not factored in. Therefore, the shortcomings of AE can be addressed by adapting it with Life Cycle (LC) concepts, which considers the whole life cycle of the production stages of the product of interest and provides a holistic score in terms of Life Cycle-Atom Economy (LC-AE).

In this chapter, we introduced the background and limitations of AE and proposed an LC-AE framework. We demonstrated the importance of LC-AE concepts via the comparison of the AE and LC-AE calculations of the four shortlisted solvents: Tetrahydrofuran (THF), 2-Methyltetrahydrofuran (2-MeTHF), *N,N*-Dimethylformamide (DMF), and Dimethyl sulfoxide (DMSO). In addition, LC-AE of the alternatives (2-MeTHF and DMSO) is investigated to find out the potentials of the alternatives in replacing the solvents (THF and DMF).

LCA is an environmental management tool that has been widely utilized in the investigation of various green and sustainable chemistry applications (refer to Chapters 2 and 5) over the years. Kalkeren *et al.* [3] stressed that green performance metrics should be expanded to include the production stages of the use of chemicals, solvents, and energy input involved to evaluate synthesis routes. Hatti-Kaul *et al.* [4] also highlighted the need for the evaluation of new products to include a life cycle perspective. Hence, in addition to LC-AE considerations, LCA will also be performed to compare the environmental performance of Production Route 5 (DMF production) and Production Route 7 (DMSO production).

2. AE and LC-AE Methodology

According to Eissen *et al.* [5], $AE_{\_product}$ can be calculated as follows:

$$AE_{product} = \frac{b_{product} \cdot MW_{product}}{a_{reactant.1} \cdot MW_{reactant.1} + \ldots + a_{reactant.m} \cdot MW_{reactant.m}} \tag{1}$$

where $b_{\_product}$, $a_{\_reactant.1}$, and $a_{\_reactant.m}$ are the stoichiometric ratios of the respective product and reactants, and $MW_{\_product}$, $MW_{\_reactant.1}$, and $MW_{\_reactant.m}$ are the molecular weight of the respective product and reactants.

As shown in Equation (1), AE calculation is commonly done within the limited boundary of a single production stage. As opposed to the LCA approach, the use of AE calculations on its own may not give a complete picture of the overall sustainability of a process or product. Therefore, a modified equation of AE calculation with life cycle concepts (LC-AE) is recommended to widen the scope of the stages of the production chain involved, based on a cradle-to-gate boundary. The LC-AE equation adapted from Eissen *et al.* [5] is presented as follows:

$$
\begin{aligned}
LC - AE(1,\ldots,n) &= \frac{b_{product} \cdot MW_{product}}{\dfrac{a_{reactant.1} \cdot MW_{reactant.1}}{LC\text{-}AE(1, \ldots, n-1)} + \sum_{j=2}^{m} \dfrac{a_{reactant.j} \cdot MW_{reactant.j}}{LC\text{-}AE(1, \ldots, n-1)}} \\
&= \frac{h}{g}
\end{aligned}
\tag{2}
$$

The difference between Equations (1) and (2) is strictly the inclusion of LC-AE(1,..., $n-1$) in Equation (2). To take into consideration LC-AE of the precedent stages, the molecular weight has to be divided by their respective underlying LC-AE(1,..., $n-1$). This equation proves useful as a more holistic investigation of the LC-AE efficiency of the chemical reactions with the aim to ensure environmental sustainability objectives by accounting for multiple intermediate steps to obtain the overall LC-AE for the final product. To test this approach, the AE and LC-AE production routes of the four selected solvents, i.e. Tetrahydrofuran (THF), 2-Methyltetrahydrofuran (2-MeTHF), N,N-Dimethylformamide (DMF), and Dimethyl sulfoxide (DMSO), are investigated.

3. Tetrahydrofuran

Projected to have a compound annual growth rate of 7.8% and estimated to value US$ 2.8 billion by the end of 2026, THF is an important chemical for major applications such as pharmaceutical, solvent and in polytetra-methylene ether glycol (PTMEG) production [6].

In the pharmaceutical sector, THF is used as a reaction medium in Grignard syntheses or lithium aluminum hydride processes. It is also widely employed in the industry as a solvent for rubber, polymer and protective coating, as well as adhesive and ink production [7].

THF has been a preferred solvent for nanoencapsulation and nanopre-cipitation processes and as an eluent in the determination of molecular weight using Gel Permeation Chromatography (GPC) [8]. THF is a pre-cursor to PTMEG, which is used as raw material for the production of fibers, such as spandex [9].

As mentioned in Chapter 2, THF can be manufactured via multiple pathways. While Chapter 2 highlights the importance of good quality LCI in producing reliable LCA results, this chapter focuses on expanding "Atom Economy" which is considered as the second principle of green chemistry. The most notable route for THF production is the Reppe pro-cess, which is the dehydrogenation of 1,4-butanediol (referred to herein as Production Route 1) [6]. Another route is butadiene acetoxylation (Production Route 2) [9].

3.1. *Production Route 1: Reppe process*

Since the 1930s, dehydration of 1,4-Butanediol (BDO) is the most com-mon production route of THF, developed by Reppe [6]. 1,4-Butynediol (ByDO) is first produced by reacting acetylene and formaldehyde, before hydrogenating to form BDO. BDO is then dehydrogenated with an acid catalyst at above 100°C to obtain THF [6]. The cradle-to-gate production route, AE, and LC-AE calculations are shown in Figure 1, Tables 1 and 2, respectively. The AE and LC-AE calculations are carried out in accor-dance with Equations (1) and (2), respectively.

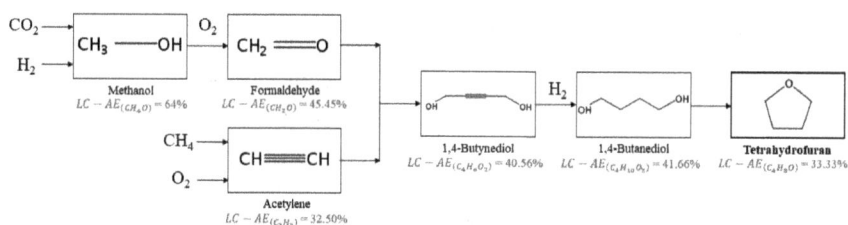

Figure 1.   Cradle-to-gate Production Route 1 for THF production.

Table 1.   AE calculation for Production Route 1.

	Tetrahydrofuran Production					
Equation	Reactants	g	Desired Product	h	Atom Economy	
$C_4H_{10}O_2 \rightarrow C_4H_8O + H_2O$	$C_4H_{10}O_2$	90	C_4H_8O	72	$AE_{(C_4H_8O)}$	80%

Table 2. LC-AE calculation for Production Route 1.

	Formaldehyde Production				
Equation	Reactants	g	Desired Product	h	Life Cycle – Atom Economy
$CO_2 + 3H_2 \rightarrow CH_4O + H_2O$	$CO_2 + 3H_2$	50	CH_4O 32		$LC-AE_{(CH_4O)}$ 64%
	$CH_4O + \dfrac{1}{2}O_2$	66	CH_2O 30		$LC-AE_{(CH_2O)}$ 45.45%

	Acetylene Production				
Equation	Reactants	g	Desired Product	h	Life Cycle – Atom Economy
$2CH_4 + \dfrac{3}{2}O_2 \rightarrow$ $C_2H_2 + 3H_2O$	$2CH_4 + \dfrac{3}{2}O_2$	80	C_2H_2 26		$LC-AE_{(C_2H_2)}$ 32.50%

	Tetrahydrofuran Production				
Equation	Reactants	g	Desired Product	h	Life Cycle – Atom Economy
$2CH_2O + C_2H_2 \rightarrow C_4H_6O_2$	$2CH_2O + C_2H_2$	212.01	$C_4H_6O_2$ 86		$LC-AE_{(C_4H_6O_2)}$ 40.56%
$C_4H_6O_2 + 2H_2 \rightarrow C_4H_{10}O_2$	$C_4H_6O_2 + 2H_2$	216.03	$C_4H_{10}O_2$ 90		$LC-AE_{(C_4H_{10}O_2)}$ 41.66%
$C_4H_{10}O_2 \rightarrow C_4H_8O + H_2O$	$C_4H_{10}O_2$	216.03	C_4H_8O 72		$LC-AE_{(C_4H_8O)}$ 33.33%

3.2. Production Route 2: Butadiene acetoxylation process

Devised by the Mitsubishi Kasei Corporation, the alternative production route of THF utilizes acetic acid and butadiene as starting materials [9]. Acetic acid is predominantly produced from carbonylation of methanol with carbon monoxide industrially [10]. Methanol is reacted catalytically with carbon monoxide at liquid phases with the use of a rhodium-based catalyst at high temperature and pressure [10].

The Houdry process, via the catalytic dehydrogenation of *n*-butane and *n*-butene, is selected for 1,3-butadiene production [11, 12]. 1,4-Diacetoxy-2-butene is obtained by oxidizing 1,3-butadiene at 3 MPa/80°C with acetic acid over a palladium-tellurium catalyst. 1,4-Diacetoxy-2-butene is then hydrogenated to 1,4-diacetoxybutane before hydrolyzing to THF [9]. The cradle-to-gate production route, AE, and LC-AE calculations are shown in Figure 2, Tables 3 and 4, respectively.

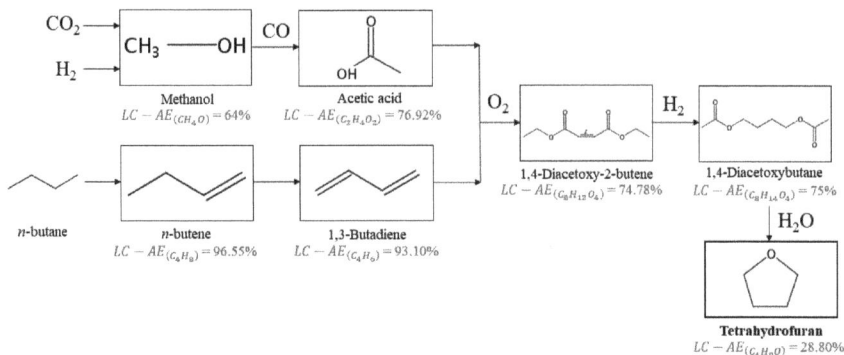

Figure 2.   Cradle-to-gate Production Route 2 for THF production.

Table 3.   AE calculation for Production Route 2.

Tetrahydrofuran Production						
Equation	Reactants	g	Desired Product	h	Atom Economy	
$C_8H_{14}O_4 + H_2O \rightarrow$ $C_4H_8O + 2C_2H_4O_2$	$C_8H_{14}O_4 + H_2O$	192	C_4H_8O	72	$AE_{(C_4H_8O)}$	37.50%

Table 4. LC-AE calculation for Production Route 2.

Acetic Acid Production

Equation	Reactants	g	Desired Product	h	Life Cycle – Atom Economy	
$CO_2 + 3H_2 \rightarrow CH_4O + H_2O$	$CO_2 + 3H_2$	50	CH_4O	32	$LC-AE_{(CH_4O)}$	64%
$CH_4O + CO \rightarrow C_2H_4O_2$	$CH_4O + CO$	78	$C_2H_4O_2$	60	$LC-AE_{(C_2H_4O_2)}$	76.92%

1,3-Butadiene Production

Equation	Reactants	g	Desired Product	h	Life Cycle – Atom Economy	
$C_4H_{10} \rightarrow C_4H_8 + H_2$	C_4H_{10}	58	C_4H_8	56	$LC-AE_{(C_4H_8)}$	96.55%
$C_4H_8 \rightarrow C_4H_6 + H_2$	C_4H_8	58	C_4H_6	54	$LC-AE_{(C_4H_6)}$	93.10%

Tetrahydrofuran Production

Equation	Reactants	g	Desired Product	h	Life Cycle – Atom Economy	
$2C_2H_4O_2 + C_4H_6 + \frac{1}{2}O_2 \rightarrow$ $C_8H_{12}O_4 + H_2O$	$2C_2H_4O_2 + C_4H_6 + 1/2\ O_2$	230.01	$C_8H_{12}O_4$	172	$LC-AE_{(C_8H_{12}O_4)}$	74.78%
$C_8H_{12}O_4 + H_2 \rightarrow C_8H_{14}O_4$	$C_8H_{12}O_4 + H_2$	232.01	$C_8H_{14}O_4$	174	$LC-AE_{(C_8H_{14}O_4)}$	75%
$C_8H_{14}O_4 + H_2O \rightarrow C_4H_8O +$ $2C_2H_4O_2$	$C_8H_{14}O_4 + H_2$	250	C_4H_8O	72	$LC-AE_{(C_4H_8O)}$	28.80%

4. Methyltetrahydrofuran

2-Methyltetrahydrofuran (2-MeTHF) is regarded as a promising, environmentally friendly alternative to THF since it is derived from renewable resources, i.e. agricultural waste [13, 14]. It has partial miscibility in water and can be a substitute for THF in applications where high miscibility with water may cause inconvenience [13]. 2-MeTHF has also been promising as an alternative solvent to THF in Gel Permeation Chromatography/ Size Exclusion Chromatography (GPC/SEC) separation [15].

Hence, 2-MeTHF is progressively used in organometallic and biphasic reactions [13]. Furthermore, 2-MeTHF can be utilized as a solvent for low-temperature lithiation, Reformatsky reaction, and metal-catalyzed coupling reactions [13].

2-MeTHF can be synthesized from renewable resources such as furfural [14, 16] or levulinic acid [17]. Furfural to 2-MeTHF follows an exhaustive hydrogenation route (Production Route 3) [14, 16] and the levulinic acid route follows cyclization and hydrogenation as the major industrial steps (Production Route 4) [17].

4.1. *Production Route 3: Synthesis from furfural*

As part of the drive for green chemistry application, bio-derived chemicals have gained much attention from both academic and industrial areas. One of the main steps in the production of bio-chemicals is the successful conversion of furfural from lignocellulosic biomass resources. Synthesis from furfural takes place by first converting xylose intermediate to furfural through dehydration of pentose using an acid catalyst, i.e. hydrochloric acid [16]. After this, furfural undergoes a series of hydrogenation cycles to obtain furfuryl alcohol and 2-methylfuran intermediates, with the aid of Cu–Zn alloy catalyst at 200–300°C [14].

Next, a nickel-based catalyst is utilized at a low temperature of 100°C to hydrogenate 2-methylfuran to yield 2-MeTHF [14]. The cradle-to-gate production route, AE, and LC-AE calculations are shown in Figure 3, Tables 5 and 6, respectively.

4.2. *Production Route 4: Cyclization and hydrogenation of levulinic acid*

Hoydonckx *et al.* used levulinic acid to produce 2-MeTHF through cyclization and hydrogenation [17]. Levulinic acid (LA) is derived from

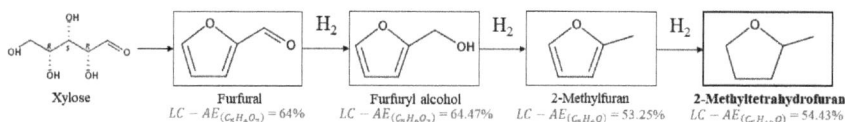

Figure 3.   Cradle-to-gate Production Route 3 for 2-MeTHF production.

Table 5.   LC-AE calculation for Production Route 3.

2-MeTHF Production						
Equation	Reactants	g	Desired Product	h	Atom Economy	
$C_5H_6O + 2H_2 \rightarrow C_5H_{10}O$	$C_5H_6O + 2H_2$	86	$C_5H_{10}O$	86	$AE_{(C_5H_{10}O)}$	100%

Table 6.   LC-AE calculation for Production Route 3.

Furfural Production					
Equation	Reactants	g	Desired Product h	Atom Economy	
$C_5H_{10}O_5 \rightarrow C_5H_4O_2 + 3H_2O$	$C_5H_{10}O_5$	150	$C_5H_4O_2$ 96	$LC-AE_{(C_5H_4O_2)}$	64%

2-MeTHF Production					
Equation	Reactants	g	Desired Product h	Atom Economy	
$C_5H_54O_2 + H_22 \rightarrow C_5H_6O_2$	$C_5H_4O_2 + H_2$	152	$C_5H_6O_2$ 98	$LC-AE_{(C_5H_6O_2)}$	64.47%
$C_5H_6O_2 + H_2 \rightarrow C_5H_6O + H_2O$	$C_5H_6O_2 + H_2$	154	C_5H_6O 82	$LC-AE_{(C_5H_6O)}$	53.25%
$C_5H_6O + 2H_2 \rightarrow C_5H_{10}O$	$C_5H_6O + 2H_2$	158.02	$C_5H_{10}O$ 86	$LC-AE_{(C_5H_{10}O)}$	54.43%

glucose wherein fructose and hydroxymethylfurfural (HMF) are the intermediates for the synthesis. Fructose is isomerized from glucose and subsequently dehydrated to form HMF and rehydrated again to form LA [18]. 3-Pentenoic acid is subsequently produced from dehydration of LA.

The intermediate reactions, i.e. 3-pentenoic acid to γ-valerolactone and γ-valerolactone to 1,4-pentanediol are catalyzed with a bimetallic catalyst in the presence of hydrogen. Finally, 2-MeTHF is obtained after dehydration of 1,4-pentanediol [19]. The cradle-to-gate production route, AE, and LC-AE calculations are shown in Figure 4, Tables 7 and 8, respectively.

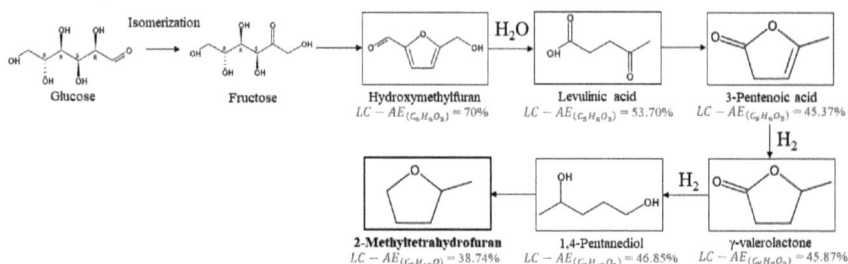

Figure 4.   Cradle-to-gate Production Route 4 for 2-MeTHF production.

Table 7.   AE calculation for Production Route 4.

2-MeTHF Production					
Equation	Reactants	g	Desired Product	h	Atom Economy
$C_5H_{12}O_2 \rightarrow C_5H_{10}O + H_2O$	$C_5H_{12}O_2$	104	$C_5H_{10}O$	86	$AE_{(C_5H_{10}O)}$ 82.69%

Table 8.   LC-AE calculation for Production Route 4.

Levulinic Acid Production					
Equation	Reactants	g	Desired Product	h	Atom Economy
$C_6H_{12}O_6 \rightarrow C_6H_6O_3 + 3H_2O$	$C_6H_{12}O_6$	180	$C_6H_6O_3$	126	$LC-AE_{(C_6H_6O_3)}$ 70%
$C_6H_6O_3 + 2H_2O \rightarrow C_5H_8O_3 + CH_2O_2$	$C_6H_6O_3 + 2H_2O$	216	$C_5H_8O_3$	116	$LC-AE_{(C_5H_8O_3)}$ 53.70%
2-MeTHF Production					
Equation	Reactants	g	Desired Product	h	Atom Economy
$C_5H_8O_3 \rightarrow C_5H_6O_2 + H_2O$	$C_5H_8O_3$	216	$C_5H_6O_2$	98	$LC-AE_{(C_5H_6O_2)}$ 45.37%

(Continued)

Table 8.   (*Continued*)

2-MeTHF Production

Equation	Reactants	g	Desired Product	h	Atom Economy	
$C_5H_6O_2 + H_2 \rightarrow C_5H_8O_2$	$C_5H_6O_2 + H_2$	218	$C_5H_8O_2$	100	$LC-AE_{(C_5H_8O_2)}$	45.87%
$C_5H_8O_2 + 2H_2 \rightarrow C_5 H_{12}O_2$	$C_5H_8O_2 + 2H_2$	222.01	$C_5H_{12}O_2$	104	$LC-AE_{(C_5H_{12}O_2)}$	46.85%
$C_5H_{12}O_2 \rightarrow C_5H_{10}O + H_2O$	$C_5H_{12}O_2$	222.03	$C_5H_{10}O$	86	$LC-AE_{(C_5H_{10}O)}$	38.74%

5. *N,N*-Dimethylformamide

N,N-Dimethylformamide (DMF) is a popular polar solvent and reagent in the realm of chemistry, which saw its importance in various chemical reactions [20]. DMF can serve as an effective ligand, dehydrating agent, reducing agent, and catalyst [20]. It is widely produced in China due to its ever-expanding synthetic leather market based on polyurethanes [21]. DMF is commercially produced via direct carbonylation of dimethyl-amine (Production Route 5) [22] or carbonylation of methanol (Production Route 6) [21].

5.1. *Production Route 5: Direct carbonylation of dimethylamine with catalyst*

DMF can be produced via direct carbonylation of dimethylamine in the presence of sodium methoxide catalyst, at an operating temperature and pressure of 120°C and 4.9 MPa [22]. The cradle-to-gate production route, AE, and LC-AE calculations are shown in Figure 5, Tables 9 and 10, respectively.

5.2. *Production Route 6: Two-step process involving carbonylation of methanol to methyl formate*

Another method for DMF production is via the two-step process involving carbonylation of methanol to methyl formate. Dimethylamine is produced by reacting methanol and ammonia at a high temperature in the presence

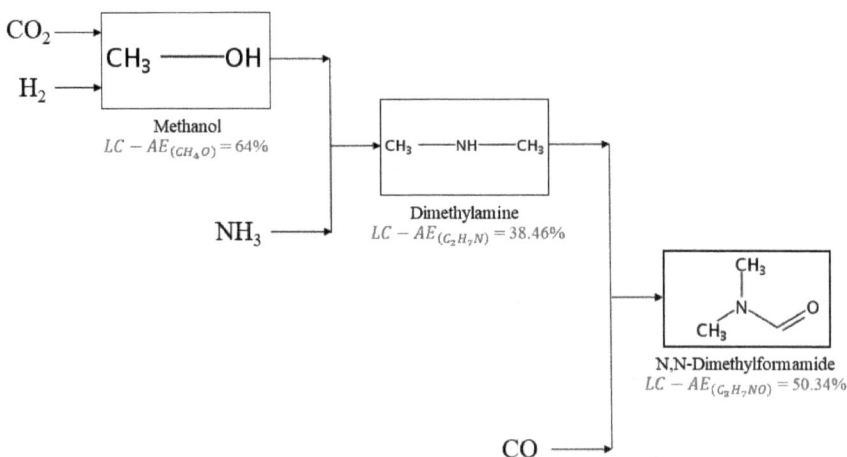

Figure 5.   Cradle-to-gate Production Route 5 for DMF production.

Table 9.   AE calculation for Production Route 5.

DMF Production						
Equation	Reactants	g	Desired Product	h	Atom Economy	
$C_2H_7N + CO \rightarrow C_3H_7NO$	$C_2H_7N + CO$	73	C_3H_7NO	73	$AE_{(C_3H_7NO)}$	100%

Table 10.   LC-AE calculation for Production Route 5.

Dimethylamine Production						
Equation	Reactants	g	Desired Product	h	Atom Economy	
$CO_2 + 3H_2 \rightarrow CH_4O + H_2O$	$CO_2 + 3H_2$	50	CH_4O	32	$LC-AE_{(CH_4O)}$	64%
$2CH_4O + NH_3 \rightarrow C_2H_7N + 2H_2O$	$2CH_4O + NH_3$	117	C_2H_7N	45	$LC-AE_{(C_2H_7N)}$	38.46%
DMF Production						
Equation	Reactants	g	Desired Product	h	Atom Economy	
$C_2H_7N + CO \rightarrow C_3H_7NO$	$C_2H_7N + CO$	145	C_3H_7NO	73	$LC-AE_{(C_3H_7NO)}$	50.34%

of a catalyst [23]. The life cycle concept is taken into account by tracing methanol back to its starting materials. In the next step, methyl formate is formed via the reaction of methanol and carbon monoxide before reacting with dimethylamine to produce DMF [21]. The Cradle-to-gate Production Route 6 for DMF production, AE, and LC-AE calculations are shown in Figure 6, Tables 11 and 12, respectively.

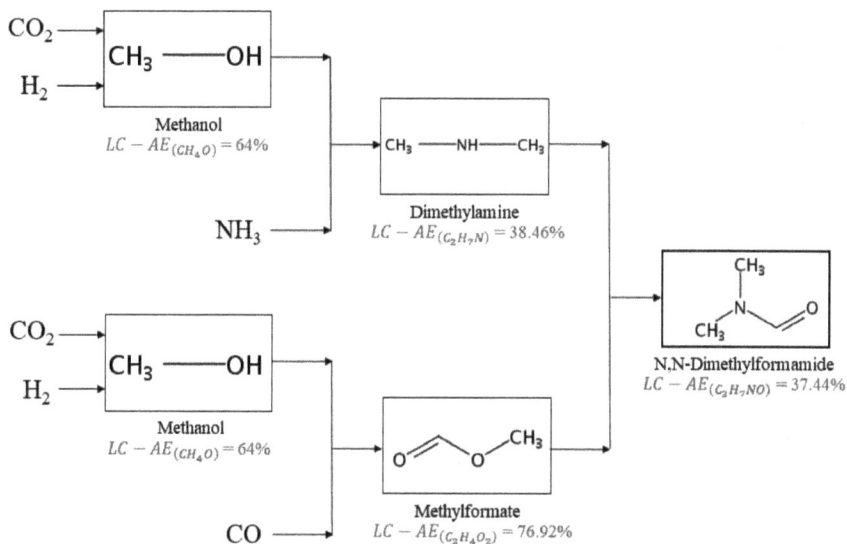

Figure 6.   Cradle-to-gate Production Route 6 for DMF production.

Table 11.   AE calculation for Production Route 6.

DMF Production					
Equation	Reactants	g	Desired Product	h	Atom Economy
$C_2H_4O_2 + C_2H_7N \rightarrow$ $C_3H_7NO + CH_4O$	$C_2H_4O_2 +$ C_2H_7N	105	C_3H_7NO	73	$LC-AE_{(C_3H_7NO)}$ 69.52%

Table 12.   LC-AE calculation for Production Route 6.

Dimethylamine Production					
Equation	Reactants	g	Desired Product	h	Atom Economy
$CO_2 + 3H_2 \rightarrow CH_4O + H_2O$	$CO_2 +$ $3H_2$	50	CH_4O	32	$LC-AE_{(CH_4O)}$ 64%

(Continued)

Table 12.   (*Continued*)

Dimethylamine Production					
Equation	Reactants	g	Desired Product	h	Atom Economy
$2CH_4O + NH_3 \rightarrow C_2H_7N + 2H_2O$	$2CH_4O + NH_3$	117	C_2H_7N	45	$LC-AE_{(C_2H_7N)}$ 38.46%

DMF Production					
Equation	Reactants	g	Desired Product	h	Atom Economy
$CH_4O + CO \rightarrow C_2H_4O_2$	$CH_4O + CO$	78	$C_2H_4O_2$	60	$LC-AE_{(C_2H_4O_2)}$ 76.92%
$C_2H_4O_2 + C_2H_7N \rightarrow C_3H_7NO + CH_4O$	$C_2H_4O_2 + C_2H_7N$	195	C_3H_7NO	73	$LC-AE_{(C_3H_7NO)}$ 37.44%

6. Dimethyl Sulfoxide

Dimethyl sulfoxide (DMSO) is a by-product of wood pulp in paper production [24]. It is employed for pharmaceutical usage and as a solvent in organic synthesis due to numerous advantages such as low cost, stability, and low toxicity [25]. DMSO can be produced by oxidation of Dimethyl sulfide (DMS), as presented in Production Route 7 [25, 26].

6.1. *Production Route 7: Oxidation of dimethyl sulfide*

DMSO is obtained via the oxidation of DMS [27]. Methanethiol is first synthesized from thiolation (reaction of hydrogen sulfide with methanol) before reacting with methanol to produce DMS [28]. In this work, hydrogen sulfide is assumed to be recycled for usage since it is typically produced as a by-product from industrial processes. Hence, its life cycle can start from itself. The cradle-to-gate production route, AE, and LC-AE calculations are shown in Figure 7, Tables 13 and 14, respectively.

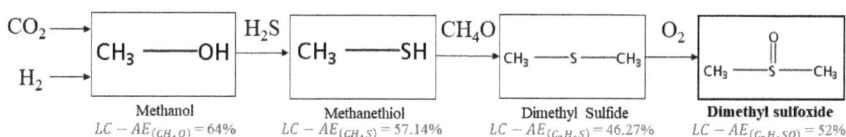

Figure 7. Cradle-to-gate Production Route 7 for DMSO production.

Table 13. AE calculation for Production Route 7.

	DMSO Production				
Equation	Reactants	g	Desired Product	h	Atom Economy
$C_2H_6S + \frac{1}{2}O_2 \rightarrow C_2H_6SO$	$C_2H_6S + \frac{1}{2}O_2$	78	C_2H_6SO	78	$AE_{(C_2H_6SO)}$ 100%

Table 14. LC-AE calculation for Production Route 7.

	DMS Production				
Equation	Reactants	g	Desired Product	h	Atom Economy
$CO_2 + 3H_2 \rightarrow CH_4O + H_2O$	$CO_2 + 3H_2$	50	CH_4O	32	$LC-AE_{(CH_4O)}$ 64%
$CH_4O + H_2S \rightarrow CH_4S + H_2O$	$CH_4O + H_2S$	84	CH_4S	48	$LC-AE_{(CH_4S)}$ 57.14%
$CH_4S + CH_4O \rightarrow C_2H_6S + H_2O$	$CH_4S + CH_4O$	134	C_2H_6S	62	$LC-AE_{(C_2H_6S)}$ 46.27%

	DMSO Production				
Equation	Reactants	g	Desired Product	h	Atom Economy
$C_2H_6S + \frac{1}{2}O_2 \rightarrow C_2H_6SO$	$C_2H_6S + \frac{1}{2}O_2$	150	C_2H_6SO	78	$LC-AE_{(C_2H_6SO)}$ 52%

7. Results and Discussion

7.1. *AE and LC-AE summarized results*

The implementation of LC-AE in all production routes shows a noticeable reduction in the percentage of overall LC-AE (%) when compared to the overall AE (%) of the same production route. This indicates the importance

Table 15.   Overall AE (%) and LC-AE (%) of Production Routes 1–7.

Production Route	End Product	Overall AE (%)	Overall LC-AE (%)
1	THF	80	33.33
2	THF	37.50	28.80
3	2-MeTHF	100	54.43
4	2-MeTHF	82.69	38.74
5	DMF	100	50.34
6	DMF	69.52	37.44
7	DMSO	100	52.00

of factoring LC-AE instead of accounting for AE alone. The overall AE (%) and LC-AE (%) of Production Routes 1–7 are depicted in Table 15. In an effort for more efficient processes, 2-MeTHF and DMSO are investigated as alternatives to THF and DMF, respectively.

For the case of 2-MeTHF, Production Routes 3 (54.43%) or 4 (38.74%) have higher LC-AE (%) compared to THF Production Routes 1 (33.33%) or 2 (28.80%). Similarly, DMSO Production Route 7 (52.00%) achieved higher LC-AE (%) compared to DMF Production Routes 5 (50.34%) or 6 (37.44%). Therefore, in comparison of a solvent to its alternative, the alternative can be chosen to substitute for the solvent since higher LC-AE efficiency is achieved.

7.2. Life Cycle Assessment

In order to provide a further overview of the potential of an alternative in substituting the respective solvent, LCA is conducted for two selected routes, Production Routes 5 and 7 to compare the environmental performance and assess the suitability of using DMSO as a substitute for DMF. A "cradle-to-gate" system boundary is employed and modeled using GaBi ts version 10.0.0.71 [29]. The cradle starts with the extraction of natural resources and the functional unit is defined as 1 kg of DMF or DMSO, respectively. The LCA model is constructed using the life cycle inventory (LCI) data from ecoinvent 3.6 [30]. ReCiPe2016 v1.1 Midpoint (H) has been selected as the life cycle impact (LCIA) model in this study, which uses impact categories that have global scope [31]. Out of the impact categories, the following seven were selected: climate change (in kg CO_2 eq.), fine particulate matter formation (in kg $PM_{2.5}$ eq.), fossil fuel depletion (in kg oil eq.), freshwater eutrophication (in kg P eq.), human toxicity (in kg 1,4-DB eq.), terrestrial acidification (in kg SO_2 eq.), and water depletion (in m^3) [29]. The overall environmental impacts of Production Routes 5 and 7 are analyzed by plotting these selected environmental impact categories into a column chart, which is displayed in Figure 8. It is worthwhile to note that Production Route 7 is a more environmentally friendly approach than Production Route 5, since all the environmental impact categories of Production Route 7 have a lower value than Production Route 5. From Table 16, Environmental impact reductions of 62.96% (climate change), 79.96% (fine particulate matter

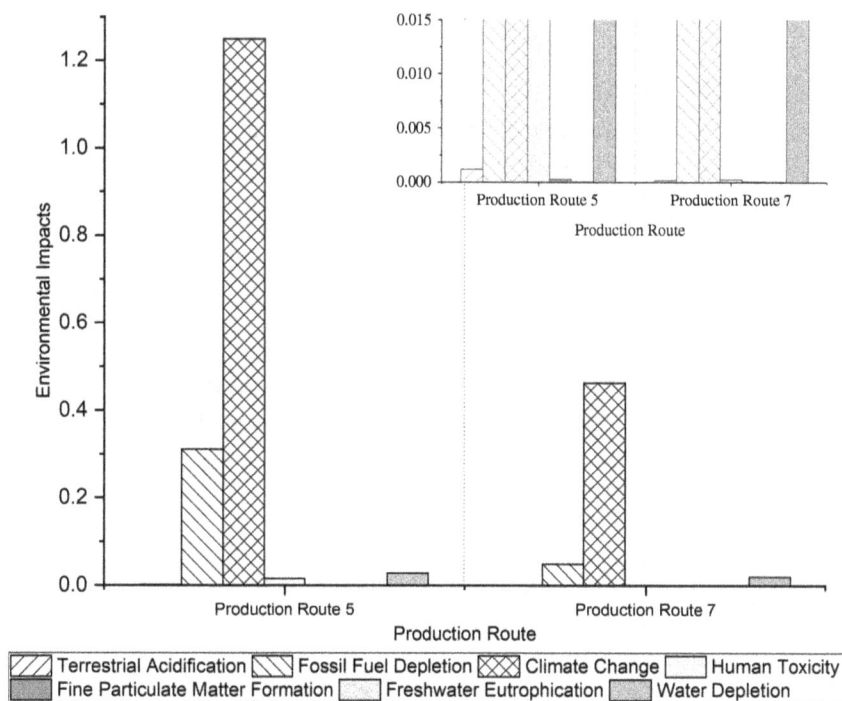

Figure 8.   Overall environmental impact results for the production of 1 kg of DMF or DMSO from Production Routes 5 and 7. The inset indicates results from 0.0 to 1.50×10^{-2} for 1 kg of DMF or DMSO.

Table 16.   Relative percentage of environmental impact categories of Production Routes 5 and 7, with Production Route 5 defined as the 100.00% column.

Environmental Impact Categories	Production Route 5	Relative Percentage (%)	Production Route 7	Relative Percentage (%)
Climate change (kg CO_2 eq.)	1.25	100	4.63×10^{-1}	37.04
Fine particulate matter formation (kg PM2.5 eq.)	2.56×10^{-4}	100	5.13×10^{-5}	20.04
Fossil fuel depletion (kg oil eq.)	3.11×10^{-1}	100	4.91×10^{-2}	15.79
Freshwater eutrophication (kg P eq.)	9.68×10^{-6}	100	8.87×10^{-6}	91.63
Human toxicity (kg 1,4-DB eq.)	1.53×10^{-2}	100	2.40×10^{-4}	1.57
Terrestrial acidification (kg SO_2 eq.)	1.20×10^{-3}	100	1.61×10^{-4}	13.40
Water depletion (m^3)	2.81×10^{-2}	100	1.98×10^{-2}	70.46

formation), 84.21% (fossil fuel depletion), 8.37% (freshwater eutrophication), 98.43% (human toxicity), 86.60% (terrestrial acidification), and 29.54% (water depletion) are achieved when production of 1 kg of DMSO is used as a substitute for 1 kg of DMF.

8. Conclusion

Among the principles of green chemistry, AE can be expressed quantitatively to depict the "degree of green" designs of products or chemical reactions. However, shortcomings of this metric exist due to its limited single-stage production scope. In order to overcome such limitations, recommendations to further expand green chemistry concepts have been suggested [32–34].

Aligned with the aim to enhance green chemistry metrics, Life Cycle-Atom Economy (LC-AE) and Life Cycle Assessment (LCA) are used as a hybrid sustainability assessment tool to investigate the environmental sustainability of THF, 2-MeTHF, DMF, and DMSO in this study. The development of the LC-AE framework shows the limitation of AE for not accounting for the raw materials production processes. The AE and LC-AE Methodology section shows the equations used in performing AE and LC-AE calculations. The calculations revealed a decreasing trend of overall LC-AE (%) from overall AE (%). LC-AE calculation indicates that the processes are indeed more inefficient than employing AE as a calculation metric. Suitable alternatives such as 2-MeTHF and DMSO are viable to replace THF and DMF, respectively, since the alternatives achieved higher LC-AE than the solvents.

The cradle-to-gate LCA of DMF (Production Route 5) and DMSO (Production Route 7) were subsequently performed and compared for the functional unit of 1 kg of DMF or DMSO. Cradle-to-gate LCA results showed that DMSO is more environmentally friendly than DMF. Environmental impact reductions of 62.96% (climate change), 79.96% (fine particulate matter formation), 84.21% (fossil fuel depletion), 8.37% (freshwater eutrophication), 98.43% (human toxicity), 86.60% (terrestrial acidification), and 29.54% (water depletion) are achieved when production of 1 kg of DMSO is used as a substitute for 1 kg of DMF.

Although useful insights can be drawn from this study, it is vital to have more resources and industrial insights relating to the solvent production processes for a more well-rounded LC-AE study. Furthermore, Life Cycle Costing (LCC) can also be included in the future to assess sustainability from an economical viewpoint.

References

[1] B. M. Trost, The atom economy — A search for synthetic efficiency. *Science* **254**(5037), 1471–1477 (1991).

[2] A. P. Dicks and A. Hent, Atom economy and reaction mass efficiency. In Sanjay K. Sharma, *Green Chemistry Metrics: A Guide to Determining and Evaluating Process Greenness.* pp. 17–44 (Springer International Publishing, Toronto, Canada, 2015).

[3] H. A. van Kalkeren, A. L. Blom, F. P. J. T. Rutjes, and M. A. J. Huijbregts, On the usefulness of life cycle assessment in early chemical methodology development: The case of organophosphorus-catalyzed Appel and Wittig reactions. *Green Chem.* **15**, 1255–1263 (2013).

[4] R. Hatti-Kaul, U. Törnvall, L. Gustafsson, and P. Börjesson, Industrial biotechnology for the production of bio-based chemicals — A cradle-to-grave perspective. *Trends Biotechnol.* **25**, 119–124 (2007).

[5] M. Eissen, R. Mazur, H.-G. Quebbemann, and K.-H. Pennemann, Atom economy and yield of synthesis sequences. *Helvetica Chimica Acta.* **87**(2), 524–535 (2004).

[6] Persistence Market Research (2018). https://www.persistencemarketresearch.com/market-research/tetrahydrofuran-market.asp.

[7] Solventis (2021). https://www.solventis.net/products/others/tetrahydrofuran/.

[8] M. E. Gindy, A. Z. Panagiotopoulos, and R. K. Prud'homme, Composite block copolymer stabilized nanoparticles: Simultaneous encapsulation of organic actives and inorganic nanostructures. *Langmuir* **24**, 83–90 (2008).

[9] H. Müller, Tetrahydrofuran. In *Ullmann's Encyclopedia of Industrial Chemistry*, Barbara Elvers (Wiley-VCH Verlag GmbH & Co. KGaA, Weinheim, Germany, 2011).

[10] G. Deshmukh and H. Manyar. Production pathways of acetic acid and its versatile applications in the food industry. In *Biotechnological Applications of Biomass*, Thalita Peixoto Basso (IntechOpen, London, United Kingdom, 2021).

[11] W. C. White, Butadiene production process overview. *Chemico-Biol. Interact.* **166**(1–3), 10–14 (2007).

[12] K. Weissermel and H. Arpe, 1,3-Diolefins. In *Industrial Organic Chemistry*, Karin Sora. pp. 107–126 (Wiley-VCH Verlag GmbH, Weinheim, Germany, 2008).

[13] D. F. Aycock, Solvent applications of 2-methyltetrahydrofuran in organometallic and biphasic reactions. *Org. Process Res. Dev.* **11**(1), 156–159 (2007).

[14] A.-G. Sicaire, M. A. Vian, A. Filly, Y. Li, A. Bily, and F. Chemat, *2-Methyltetrahydrofuran: Main Properties, Production Processes, and Application in Extraction of Natural Products.* pp. 253–268 (Springer, Berlin, Heidelberg, 2014).

[15] Greener GPC/SEC (2021). https://analyticalscience.wiley.com/do/10.1002/was.00170054.

[16] V. Choudhary, A. B. Pinar, S. I. Sandler, D. G. Vlachos, and R. F. Lobo, Conversion of xylose to furfural using lewis and brønsted acid catalysts in aqueous media. *ACS Catal.* **1**, 1724 (2011).

[17] H. E. Hoydonckx, W. M. Van Rhijn, W. Van Rhijn, D. E. De Vos, and P. A. Jacobs, Furfural and derivatives. In *Ullmann's Encyclopedia of Industrial Chemistry*, Barbara Elvers (Wiley-VCH Verlag GmbH & Co. KGaA, Weinheim, Germany, 2007).

[18] K. Kumar, F. Parveen, T. Patra, and S. Upadhyayula, Hydrothermal conversion of glucose to levulinic acid using multifunctional ionic liquids: Effects of metal ion co-catalysts on the product yield. *New J. Chem.* **42**(1), 228–236 (2018).

[19] D. C. Elliott and J. G. Frye, Hydrogenated 5-carbon compound and method of making. US Patent US5883266A (January, 1998).

[20] S. Ding and N. Jiao, N,N-dimethylformamide: A multipurpose building block. *Angewandte Chemie — Int. Ed.* **51**(37), 9226–9237 (2012).

[21] J. A. Marsella, Dimethylformamide. In *Kirk-Othmer Encyclopedia of Chemical Technology*, WATCHER. pp. 1–9 (John Wiley & Sons, Inc., New York, United States, 2013).

[22] Nitto Chemical Industry Co., Ltd. Brit. Patent 1213173 (November, 1970).

[23] J. Dick and C. Rader, 12.57 Dimethyl Terephthalate. In *Raw Materials Supply Chain for Rubber Products — Overview of the Global Use of Raw Materials, Polymers, Compounding Ingredients, and Chemical Intermediates*, Cheryl Hamilton, p. 592 (Hanser Publishers, Munich, Germany, 2014).

[24] K. Capriotti and J. A. Capriotti, Dimethyl sulfoxide: History, chemistry, and clinical utility in dermatology. *J. Clin. Aesthetic Dermatol.* **5**(9), 24–26 (2012).

[25] J.-C. Xiang, Q.-H. Gao, and A.-X. Wu, 7.1 a brief introduction of DMSO. In *Solvents as Reagents in Organic Synthesis — Reactions and Applications*, Xiao-Feng Wu (John Wiley & Sons, Weinheim, Germany, 2018).

[26] K.-M. Roy, Sulfones and sulfoxides. In *Ullmann's Encyclopedia of Industrial Chemistry*, Barbara Elvers (Wiley-VCH Verlag GmbH & Co. KGaA, Weinheim, Germany, 2000).

[27] M. V. Gavrilin, G. V. Sen'chukova, and E. V. Kompantseva, Methods for the synthesis and analysis of dimethyl sulfoxide (a review). *Pharma. Chem. J.* **34**(9), 490–493 (2000).

[28] A. V. Mashkina, Synthesis of methyl mercaptan from methanol and hydrogen sulfide at elevated pressure on an industrial catalyst. *Petroleum Chem.* **46**(1), 28–33 (2006).

[29] GaBi ts, version 10.0.0.71. thinkstep GmbH, Sphera. (Leinfelden-Echterdingen, Germany, 2021).

[30] Ecoinvent 3.6 (2021). https://www.ecoinvent.org/database/database.html.

[31] M. A. J. Huijbregts, Z. J. N. Steinmann, P. M. F. Elshout, G. Stam, F. Verones, M. Vieira *et al.*, ReCiPe2016: A harmonised life cycle impact assessment method at midpoint and endpoint level. *Int. J. Life Cycle Assess.* **22**, 138–147 (2017).

[32] P. T. Anastas and M. M. Kirchhoff, Origins, current status, and future challenges of green chemistry. *Acc. Chem. Res.* **35**, 686–694 (2002).

[33] W. Wang, J. Lü, L. Zhang, and Z. Li, Real atom economy and its application for evaluation the green degree of a process. *Front. Chem. Sci. Eng.* **5**, 349–354 (2011).

[34] G. J. Ruiz-Mercado, M. A. Gonzalez, and R. L. Smith, Expanding GREENSCOPE beyond the gate: A green chemistry and life cycle perspective. *Clean Tech. Environ. Policy* **16**, 703–717 (2014).

https://doi.org/10.1142/9789811245800_0007

Chapter 7

Gate-to-Gate Life Cycle Assessment of Solid Waste Conversion of Black Aluminum Dross to γ-Alumina as Catalyst Support for Biofuel Production

Nor Adilla RASHIDI[*,‡], Yee Ho CHAI[*,§],
Suzana YUSUP[*,¶], Sivapalan KATHIRAVALE[†,‖] and
Noor Mohd Syeqqal ISMAIL[†,**]

[*]*Biomass Processing Laboratory, HICoE — Centre for Biofuel and Biochemical Research, Institute of Self-Sustainable Building, Department of Chemical Engineering, Universiti Teknologi PETRONAS, 32610, Seri Iskandar, Perak, Malaysia*

[†]*Environmental Preservation and Innovation Centre Sdn. Bhd., CENVIRO Eco-park Lot PT 8436, Mukim Jimah Ladang Tanah Merah A3 Division 71960, Taman Bukit Pelandok, 71960 Port Dickson, Negeri Sembilan, Malaysia*

[‡]*adilla.rashidi@utp.edu.my*

[§]*yeeho.chai@utp.edu.my*

[¶]*drsuzana_yusuf@utp.edu.my*

[‖]*Sivapalan@epic.org.my*

[**]*Syeqqal@epic.org.my*

Aluminum is remarkably in demand due to its wider applicability in various applications. Nevertheless, the dross residues need to be treated prior to discharge, not only to recover the valuable metals but also to assess potential economic and environmental impacts. The present study describes the gate-to-gate life cycle assessment of the aluminum recovery from dross wastes via hydrolysis–leaching–calcination technology. Two scenarios that include dual acid leaching techniques using hydrochloric acid (HCl) and sulfuric acid (H_2SO_4) and utilization of non-renewable-based electricity and renewable-derived electricity impacts were explored. Further, the influence of both scenarios on the Global Warming Potential (GWP) and Marine Aquatic Ecotoxicity Potential (MAETP) was also analyzed. Sensitivity analysis with regard to variations in leaching efficiency (30%, 50%, and 70%) showed good improvement in acidification (36.4% improvement) and eutrophication (43.5% improvement) categories, while other environmental categories are insignificant. Overall, this chapter highlights the conversion methods of black aluminum dross to gamma alumina and the life cycle assessment for 1 ton production of γ-alumina derived from black aluminum dross.

1. Introduction

Due to the unique combination properties of aluminum and its alloy (i.e. lightweight, low cost, high durability and ductility, good thermal conductivities, and high corrosion resistance) [1, 2], the market of such commodities has shown an acceleration trend due to its extensive applications such as in transportation and in building/construction sector. Watari *et al.* [3] reported that the metal industry will be monopolized by aluminum commodity by 2050 (with a growth rate of 215% relative to the 2010 level), as compared to other metals such as copper, nickel, iron, zinc, and lead.

Regardless of the blooming aluminum industry, this sector causes severe environmental burden due to the massive production of solid wastes — aluminum dross. Specifically, Sultana *et al.* [4] reported that 20–25 kg of dross is produced per metric ton of molten aluminum and a total of 69 million tons of dross was recorded in 2019 [5]. The dross consists of hazardous metal ions (i.e. alumina, silica, magnesia) and is often disposed of through landfilling in the vicinity of factories, whereupon it will result in leaching problems and cause soil/groundwater pollution, in addition to the emission of noxious odors and harmful/flammable gases such as methane (CH_4), hydrogen sulfide (H_2S), and ammonia (NH_3) upon contact with water [6–8]. Similarly in Malaysia, the aluminum dross that has been classified as a toxic solid waste was subjected to landfilling, which costs around RM 2,000 per ton of dross.

Thereby, with respect to the high cost of waste management of aluminum dross, conversion of this industrial waste (particularly black dross or commonly known as the secondary dross, lean dross, or dry dross) to a value-added material is a promising option. Herein, black dross conversion to useful materials is given utmost consideration due to the following justifications:

- *High production rate*: It has been projected that the production rate of white dross (other names: primary dross, rich dross, or wet dross) and black dross is 1.5–2.5% and 8–15% per ton of molten aluminum produced [9], respectively.
- *Solid waste management*: It is reported that only the white dross is subjected to recycling treatment and aluminum extraction, whereas the generated black dross is sent to landfills [10, 11] due to lower aluminum contents (5–20%).
- *Physiochemical characteristics*: Nduka *et al.* [12] reported that the black dross has higher salt content and gas evolution as compared to the white dross. Thus, the black dross has been classified as hazardous/toxic by the European Union (code: 100309) [7, 13].

Hence, with regard to the considerable amounts of aluminum content in the black dross, this industrial waste can be valorized for γ-alumina production. γ-alumina production is significant due to its extensive usage as an adsorbent or catalyst support, attributed to its low cost, high porosity, pore size, and BET surface area, in addition to high thermochemical stability [14, 15]. Further, conversion of black dross to γ-alumina helps contribute to the following: (i) environmental protection as the industrial waste can be reduced and (ii) economic sector in terms of reduction of disposal costs associated with black dross landfilling, and inexpensive γ-alumina production can be attained, as the industrial waste is used as the starting materials, as opposed to the conventional approach that depends on pricey synthetic chemicals as the starting materials, such as aluminum nitrate nonahydrate, aluminum isopropoxide, and aluminum ammonium carbonate hydroxide [16, 17].

Overall, this chapter focuses on gate-to-gate life cycle assessment (LCA) of black dross conversion to γ-alumina. To the best of our knowledge, LCA studies on black dross conversion to γ-alumina have not been presented yet in literature; therefore, it implies the significance of this chapter as stated by Zhu *et al.* [18] that LCA studies related to the aluminum dross conversion are rather limited in the literature.

2. Case Study

2.1. *Process descriptions*

In this study, only chemical reactions that involved aluminum-based compounds were taken into consideration. This is also in agreement with Meshram *et al.* [19] who reported that the dross consists of aluminum and alumina (aluminum oxide), with the former embedded into the latter's structure. On top of that, Table 1 shows that the dross primarily contains the following: −87% aluminum-based compounds (Al_{metal}, Al_2O_3, AlN), −1.5% SiO_2, and oxides of Na, Mg, K, and Ca.

Table 1. Chemical properties of aluminum dross.

Chemical Properties	Fraction (%)
Al_2O_3	78.24
AlN	7.23
Al_{metal}	1.16
SiO_2	1.46
Na_2O	6.28
CaO	0.47
TiO_2	0.13
MgO	1.97
K_2O	1.47
Zn	0.05
Cu	0.51
Mn	0.03
Fe	1.00

Source: Modified from Ref. [20].

2.1.1. *Water washing*

The as-received aluminum dross is first subjected to washing treatment. This is done to promote dross dissolution during the subsequent acidic leaching. Herein, data for the optimum water washing process is referred from Feng *et al.* [21] where the optimal operating conditions are as follows: −102°C for 4 hrs. The reactions involved in the process are described in Equations (1)–(3). However, it should be noted that the composition of AIP and Al_4C_3 in the dross is insignificant [22] and thus were omitted from the analysis in this study:

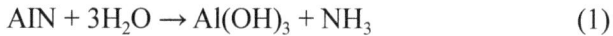

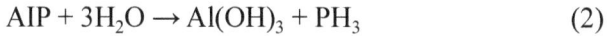

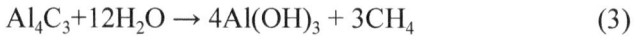

$$AIN + 3H_2O \rightarrow Al(OH)_3 + NH_3 \tag{1}$$

$$AIP + 3H_2O \rightarrow Al(OH)_3 + PH_3 \tag{2}$$

$$Al_4C_3 + 12H_2O \rightarrow 4Al(OH)_3 + 3CH_4 \tag{3}$$

2.1.2. *Leaching process*

Prior to the acidic leaching process, the pretreated dross was filtered and subjected to the drying process to remove the free water from the dross. Subsequently, the dried dross reacted with a certain amount of acids, i.e. H_2SO_4 and HCl under a constant temperature of 70°C following Feng *et al.* [21]. Upon completion of the leaching process, the purified dross was separated from the leach liquor containing aluminum sulfate/ aluminum chloride, after which the liquor was subjected to the subsequent precipitation stage. The main reactions during the H_2SO_4 and HCl leaching are as shown in Equations (4)–(8) [21, 23]. Irrespective of the types of acid used during the leaching process, a considerable amount of hydrogen was produced:

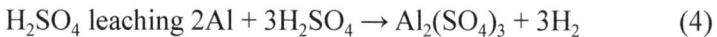

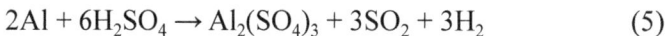

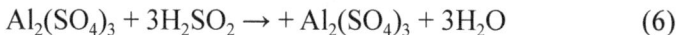

$$H_2SO_4 \text{ leaching } 2Al + 3H_2SO_4 \rightarrow Al_2(SO_4)_3 + 3H_2 \tag{4}$$

$$2Al + 6H_2SO_4 \rightarrow Al_2(SO_4)_3 + 3SO_2 + 3H_2 \tag{5}$$

$$Al_2(SO_4)_3 + 3H_2SO_2 \rightarrow + Al_2(SO_4)_3 + 3H_2O \tag{6}$$

$$\text{HCl leaching } Al + 3HCl \rightarrow AlCl_3 + 1.5H_2 \tag{7}$$

$$Al_2O_3 + 6HCl \rightarrow 2AlCl_3 + 3H_2O \tag{8}$$

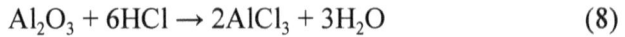

2.1.3. Precipitation and post-treatment

The precipitation stage was carried out by adding a considerable amount of sodium hydroxide to the aluminum sulfate/aluminum chloride leachate, following the stoichiometric reaction shown in Equations (9) and (10) for the case of H_2SO_4 and HCl leaching, respectively. Prior to the subsequent calcination, the precipitates were pre-dried at a temperature of 80°C for 2 hours to remove the excess water:

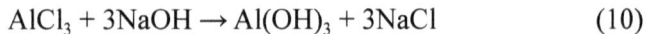

$$Al_2(SO_4)_3 + 6NaOH \rightarrow 2AlOH_3 + 3Na_2SO_4 \tag{9}$$

$$AlCl_3 + 3NaOH \rightarrow Al(OH)_3 + 3NaCl \tag{10}$$

2.1.4. Calcination process

The calcination process of the dried precipitate was carried out in a muffle furnace at a temperature of 800°C (10°C/min) for 2 hours [24], after which the muffle furnace was cooled down to room temperature. The reaction during this final stage is as follows [25]:

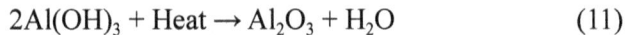

$$2Al(OH)_3 + Heat \rightarrow Al_2O_3 + H_2O \tag{11}$$

3. Life Cycle Assessment

The LCA modeling procedure was carried out according to the ISO 14040-14044 standards: (i) goal and scope, (ii) life cycle inventory, (iii) impact assessment, and (iv) result interpretation.

3.1. Goal and scope of the study

The LCA scope in this study refers to the gate-to-gate boundary for the functional unit of 1,000 kg γ-alumina production. The LCA model comprised four stages: water washing (pre-treatment), acid leaching,

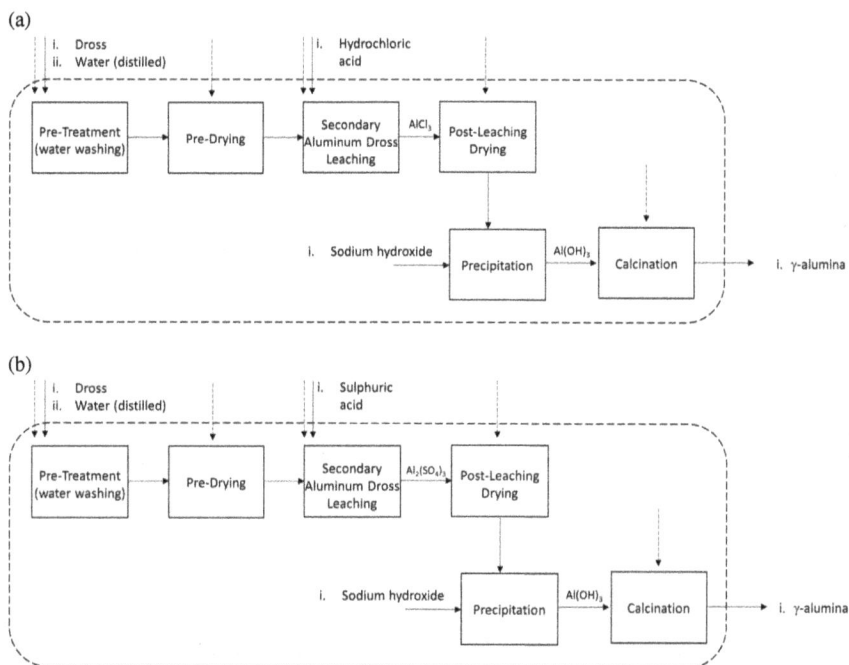

Figure 1.   System boundary defined within the dashed box illustrates the production of 1,000 kg γ-alumina from black aluminum dross via (a) HCl leaching and (b) H_2SO_4 leaching. Chemical and energy inputs are represented by solid and dashed arrows, respectively.

precipitation, and finally calcination process. In this analysis, acid leaching using H_2SO_4 and HCl has been emphasized due to having good efficiency in solubilizing the alumina compounds and insolubility of silica in an acidic medium. This is opposite to the alkaline route that favorably aids the dissolution of the silica phase, which later brings about a major concern for silica removal, prior to the precipitation of aluminum hydroxide ($Al(OH)_3$) [26]. Gabi software (Version 10.0.1.92) was used to model the LCA system and generate the associated set of environmental impacts. The database platform of Professional Database version 2021 was applied; all input–output data were allocated to the functional unit of 1 ton of γ-alumina production. The LCA system boundary is described in Figure 1.

3.2. *Life cycle inventory*

An inventory analysis was conducted to quantify and track the inputs and outputs of the materials and energy flows within the system boundary. As there are no established data about the γ-alumina production from black aluminum dross at an industrial scale, information about the conversion of aluminum-based compounds residing within the dross are obtained from various literature studies. Subsequently, a simplified gate-to-gate analysis is modeled to estimate the overall materials and energy flows required and produced.

The operating conditions and details of individual unit operations such as pre- and post-treatments, leaching, precipitation, and calcination processes are derived from various literature data, as readily elaborated in Section 2. The experimental conditions for HCl- and H_2SO_4-based conversion process are outlined in Table 2.

3.3. *Life cycle impact assessment*

According to the International Standard for LCA, the absolute life cycle impact assessment (LCIA) method selected to be used is not specified and is dependent on the type of study [28]. Nevertheless, it is noteworthy that the latest LCIA methodology must always be used. This study considers the use of the CML 2001 (January 2016 version) method that determines specific impact indicators with respect to the emissions from each individual processing stage defined in the system boundary [29]. Geographically, the CML 2001 method considers the environmental impacts at a global scale when compared to TRACI (North America) and ReCiPe (Europe) [30].

The environmental impact categories considered in this study include the following: global warming potential (GWP), acidification potential (AP), eutrophication potential (EP), human toxicity potential (HTP), and marine aquatic ecotoxicity potential (MAEP). Furthermore, assessment of inorganic emissions such as SO_x and NH_3 as one of the environmental impacts has been included. This is partly attributed to the use of H_2SO_4 in the leaching process and potential NH_3 gas generated from the water washing.

Table 2. LCI of secondary aluminum dross conversion to 1,000 kg γ-alumina with 30% HCl and H_2SO_4 acid leaching efficiency.

Parameters	Amounts	Remarks	Assumptions	References
Pre-treatment water washing				
Electricity	11,716.69 MJ	Electricity energy required to heat up water from 25°C to 102°C.	No heat loss was considered.	[21]
Deionized water	9,057.43 kg	Heat capacity of water: 4.164 J/g/°C	Ratio of 1:3 dross-to-water	[27]
Aluminum hydroxide	415.39 kg	Intermediary by-product produced from pre-treatment washing following Equation (1)		—
Post-washing drying				
Electricity	836.90 MJ	Electricity energy required to remove 80% of water from wetted aluminum dross.	Total drying duration of 8 hours was considered.	—
Acid leaching				
Electricity	947.15 MJ[a] 1,274.01 MJ[b]	Electricity energy required to heat up acid solution from 25°C to 70°C.	Total leaching duration of 1 hour was considered.	—
Hydrochloric acid	5,209.85 kg	Acid leaching follows Equations (7) and (8).	Aluminum-based compounds and metals were fully leached.	—
Sulfuric acid	7,007.75 kg	Acid leaching follows Equations (9) and (10).		
Post-leaching drying				
Electricity	6,859.54 MJ[a] 2,478.45 MJ[b]	Electricity energy required to completely remove water content from aluminum salts.	Total drying duration of 2 hours was considered. Assume 20% water moisture content retained after filtration from the acid leaching process.	—

(Continued)

Table 2.   (*Continued*)

Parameters	Amounts	Remarks	Assumptions	References
Aluminum chloride	1,905.43 kg[a]	Production of intermediate product from aluminum dross acid leaching.	—	—
Aluminum sulfate	2,444.66 kg[b]			
Precipitation				
Sodium hydroxide	1,714.67 kg	The precipitation process follows Equation (10).	Aluminum chloride was fully precipitated.	—
Aluminum hydroxide	1,114.62 kg	Aluminum hydroxide produced.		
Calcination				
Electricity	275,402.10 MJ	Electricity energy required to calcine aluminum hydroxide at 700°C for 2 hours.	Scaling of 4.5 kW rating for 180 g lab-scale equipment was considered.	—
γ-Alumina	1,000.00 kg	γ-Alumina produced.		

Note: [a] Refers to HCl acid leaching process.
[b] Refers to H_2SO_4 acid leaching process.

4. Results and Discussion

4.1. *Interpretation of life cycle assessment results*

The conversion of black aluminum dross to γ-alumina via the acidic leaching process is a very energy-intensive process. This is in agreement with Reddy and Neeraja [31] who previously reported that the aluminum recovery from dross is an energy-intensive process, where the electricity consumption accounts for 74% of the total energy required. As such, electricity from the grid is one of the major inputs (see Table 3) within the system boundary that significantly contributes to the GWP. In addition, the reaction process that releases significant amounts of organic and inorganic emissions has large detrimental effects on the marine ecosystem.

As shown in Table 3, for every 1,000 kg of γ-alumina produced, the calcination process contributes a significant portion to GWP with approximately 36,000 kg CO_2 eq. emission in the HCl leaching process, due to its intrinsic intensive electricity consumption. Electricity contributes more than 90% of the total CO_2 eq. emission, cumulative from all unit processes defined in the gate-to-gate boundary system. It is estimated that nearly 40.9 kg CO_2 eq. is released for every kg of γ-alumina produced via the HCl leaching process, while the utilization of H_2SO_4 acid for the leaching process reduced the emission to 38.3 kg CO_2 eq. instead. Further, electricity contributes 93.2% (37.9 kg) of the total CO_2 equivalent emission of the entire unit process when HCl acid was used, while the remaining major contributors are derived from HCl (2.9%) and NaOH (3.8%) utilization. Compared to the HCl acid leaching process, electricity in the H_2SO_4 acid leaching process contributes up to 99.7% (38.1 kg) of total CO_2 eq. emission, while the H_2SO_4 acid production process accounts for -1.4 kg CO_2 eq. emission (-3.7%) instead for every kg of γ-alumina produced.

The acidification potential that is expressed in terms of kilogram of sulfur dioxide (SO_2) equivalence represents the potential of that process to contribute to the acid rain phenomenon. It is worthwhile to mention that the H_2SO_4 leaching process leads to higher acidification potential when compared against HCl (58.43 *vs.* 1.43), given the reaction leads to SO_2 production (Equation (5)). Moreover, it has been reported that electricity consumption can also contribute to acidification potential [32].

The use of electricity from natural gas sources and contributions from the production of chemical feedstocks contributed to more than 350 kg of dichlorobenzene (DCB) eq. emission in marine aquatic ecotoxicity

Table 3. LCIA contributions calculated from different acid leaching processes for γ-alumina production to the environmental impacts.

Stages in Industrial-Scale Conversion of Black Aluminum Dross to 1,000 kg γ-Alumina via HCl Acid Leaching Process

Impact Category	Unit	Electricity for Water Washing	Electricity for Post-Washing Drying	Electricity for Acid Leaching Process	Electricity for Post-Leaching Drying	Electricity for Calcination	HCl	Sodium Hydroxide	Water	Total
GWP	kg CO_2 eq.	1,532.44	109.46	123.88	191.70	36,020.10	1,293.25	1752.84	12.94	40,912.73
AP	kg SO_2 eq.	0.67	0.05	0.05	0.08	15.8	1.43	3.33	0.02	21.43
EP	kg PO_4^{3-} eq.	0.15	0.01	0.01	0.02	3.41	0.30	0.56	0.01	4.47
HTP	kg DCB eq.	23.51	1.68	1.90	2.94	552.64	47.37	54.34	0.50	684.88
MAETP	kg DCB eq.	9,083.45	648.82	734.29	1,136.31	213,507.40	56,132.84	122,300.90	955.26	404,499.27
SO_x	Kg	—	—	—	—	—	—	—	—	4.52
NH_3	Kg	—	—	—	—	—	—	—	—	30.00

Stages in Industrial-scale Conversion of Black Aluminum Dross to 1,000 kg γ-Alumina via H_2SO_4 Acid Leaching Process

Impact Category	Unit	Electricity for Pre-Treatment	Electricity for Pre-Treatment Drying	Electricity for Acid Leaching Process	Electricity for Post-Leaching Drying	Electricity for Calcination	H_2SO_4	Sodium Hydroxide	Water	Total
GWP	kg CO_2 eq.	1,530.96	109.35	166.47	323.85	35,985.40	−1,408.59	1,529.64	11.25	38,248.33
AP	kg SO_2 eq.	0.67	0.05	0.07	0.14	15.80	58.43	3.33	0.02	78.51
EP	kg PO_4^{3-} eq.	0.15	0.01	0.02	0.03	3.41	−0.11	0.56	0.01	4.08
HTP	kg DCB eq.	23.54	1.68	2.56	4.97	552.64	−12.64	54.38	0.50	627.63
MAETP	kg DCB eq.	9,083.45	648.82	987.68	1,921.43	213,507.40	9,015.05	122,300.9	955.26	358,419.99
SO_x	Kg	—	—	—	—	—	—	—	—	51.43
NH_3	Kg	—	—	—	—	—	—	—	—	30.00

potential (MAETP) for every kg of γ-alumina produced. By employing weak point analysis with a 30% threshold, the significant use of electricity was identified as a hotspot in contributing to the exorbitant values of DCB eq. emission. This is hugely attributed to the large emission of the industrial by-product hydrogen fluoride derived from the heavy utilization of electricity contributing to an average of 92.9% of the total DCB eq. emission. Hence, renewable energy sources, such as hydropower or biogas, should be considered to mitigate the detrimental effects of hydrogen fluoride emissions on the marine aquatic ecosystem. The use of potential renewable energy sources will be further discussed in our sensitivity analysis.

In addition, marine aquatic ecotoxicity and human toxicity potential environmental impacts, both expressed as kg DCB eq., are similar to one another. Evident differences are found in the contributing pathways to produce HCl and H_2SO_4, where the former exhibits greater toxicity to both environmental impacts. This is largely attributed to the stronger dissociation of H^+ to form strong Cl^- as compared to a weaker SO_4^- acid base.

Overall, the large emission of CO_2 from both processes is apparent and the deployment of carbon capture and storage (CCS) process should be emphasized to ensure the production of γ-alumina is sustainable with minimal environmental impact. Furthermore, the use of strong acids in leaching of aluminum dross favors acidification but harms the marine aquatic ecosystem significantly. Thus, more research should be conducted to ensure a more sustainable and viable solution is available.

The comparison of the environmental performance of each individual process unit in the HCl and H_2SO_4 process for alumina recovery from secondary dross is summarized in Table 3.

4.2. Sensitivity analysis

4.2.1. Acid leaching efficiency

The leaching of secondary aluminum dross is an important contributor to the sensitivity analysis. The consumption of raw feedstock materials and energy demand requirements are highly dependent on the leaching efficiency of aluminum dross feedstock. In other words, the acidic leaching process is the rate-determining step in the entire production system of γ-alumina, which holistically affects the entire environmental impact assessment.

Figure 2 illustrates the performance indicator of each respective environmental impact with respect to 30%, 50%, and 70% acidic leaching efficiencies, respectively. GWP in the HCl acid leaching process shows marginal improvements with a reduction of 2.2% when leaching efficiency increases from 30% to 70%, while the H_2SO_4 leaching process shows no improvement. Similarly, both HCl and H_2SO_4 acid leaching efficiencies have minor effects on human toxicity potential, marine aquatic ecotoxicity potential, and acidification potential (except for the H_2SO_4 leaching process). Eutrophication potential is 43.6% because of falling inorganic NH_3 emissions into soil for both processes. Acidification potential reduced by 36.4% at higher H_2SO_4 leaching efficiency due to the significant reduction of SO_x gas emissions, namely sulfuric dioxide and sulfuric trioxide, into air. This is partly attributed to a lesser generation of SO_x emissions as a result of a more complete aluminum dross leaching reaction.

4.2.2. *Electricity (natural gas) source replacement*

Two new scenarios based on renewable sources such as hydropower and biogas are proposed. Table 4 shows the changes in the performance indicators of the environmental impact categories as discussed above.

When hydropower is utilized as a source for energy generation, the performance indicators of all environmental impacts, except for human toxicity, decrease significantly. Global warming effect declines by more than 90% for both the acid leaching processes, as it limits the emission of CO_2 through the elimination of non-renewable natural gas consumption. Similarly, both acidification and eutrophication impacts decrease, given that SO_x emissions are mainly contributed from natural gas-based electricity utilization. The transition to hydropower also sees a decrease in marine aquatic ecotoxicity levels by more than 50%, with the NaOH production process being identified as the hotspot instead of electricity generation. Nevertheless, the heavy use of river water for the generation of hydropower sees an increase in the emission of polychlorinated dibenzofurans (PCDFs) found in the wastewater discharges. There are 17 congeners of PCDFs which are toxic and tend to accumulate in the organs of marine organisms [33], thus inflicting greater impacts in terms of human toxicity.

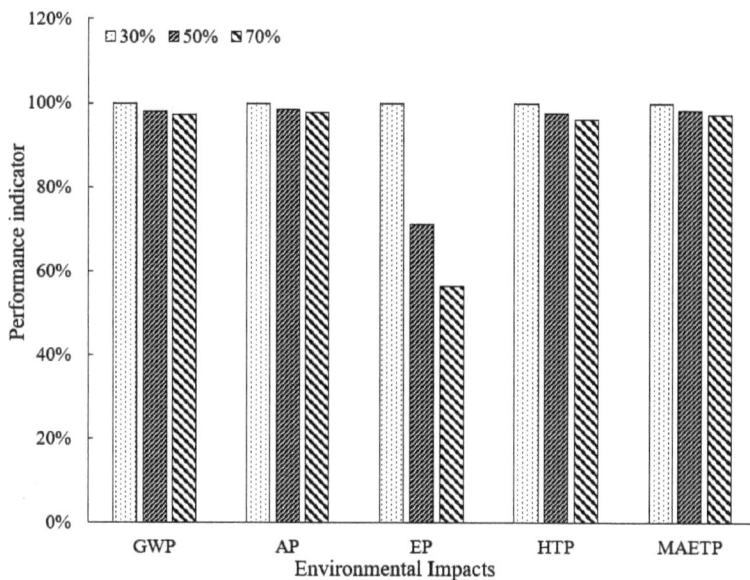

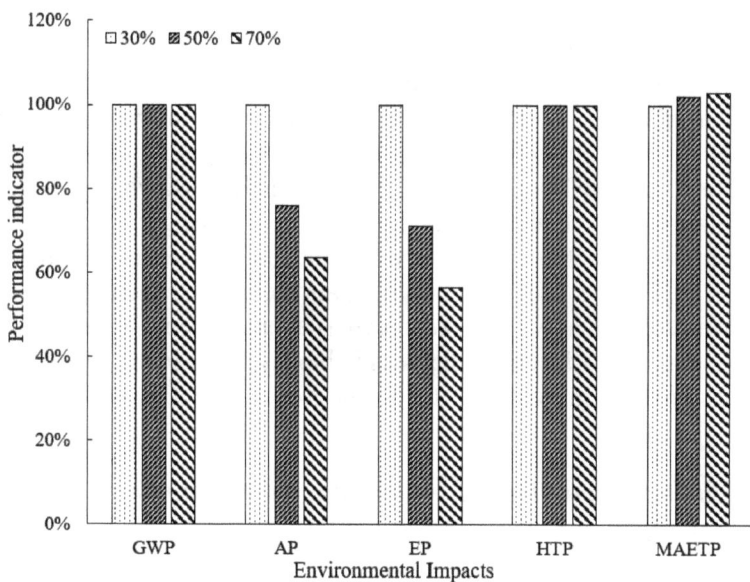

Figure 2. Performance indicators of individual environmental impacts based on 30%, 50%, and 70% HCl acid leaching (top) and H_2SO_4 leaching process (bottom), respectively.

Table 4.   Changes in performance indicators in new scenarios.

| Environmental Impacts | Changes in Performance Indicators (%) | | | |
| | HCl Leaching | | H$_2$SO$_4$ Leaching | |
	HydroPower	Biogas	HydroPower	Biogas
GWP	−92.03	−49.17	−98.34	−52.54
AP	−75.07	1403.62	−20.60	385.18
EP	−9.73	177.08	−9.88	179.83
HTP	8.94	188.65	9.80	206.80
MAETP	−50.41	71.08	−57.14	80.56

However, the transition from natural gas to biogas only sees improvement in terms of the GWP, where at least 38,255 kg CO$_2$ eq. emission is reduced to 20,685 kg CO$_2$ eq. emission for every 1,000 kg of γ-alumina production. Instead of CO$_2$ and SO$_x$ emissions, biogas contains large amounts of hydrogen sulfide that is acidic and corrosive intrinsically. Furthermore, hydrogen sulfide can potentially oxidize to SO$_2$ and H$_2$SO$_4$ [34], which can further contribute to acid rains [35], hence making the transition of energy sources redundant. Similarly, nitrogen compounds (nitrates) and benzenes are often associated with biogas production. When available in a large concentration, nitrates easily dissolve in water and accumulate excessive nutrients that can promote algae growth [36], however, benzene is a hazardous air pollutant that possesses carcinogenic properties [37]. The utilization of biogas also sees a detrimental impact on the marine ecosystem, with nickel metal as the main contributor. Nickel is often used as the co-factor of enzymes in biogas plants to avoid a decrease in biogas production over an extended period of time [38].

4.3. *Further discussions: Limitations and assumptions of the life cycle assessment model*

The system boundary in this study considered a "gate-to-gate" analysis which began with the pre-treatment of the collected black aluminum dross (feedstock) and ended with the production of γ-alumina (end product). The consideration of black aluminum dross collection, segregation, and transportation effects is excluded, while the effects of pre- and post-treatment and their subsequent reaction processes are focused on this study. Similarly, energy flows required to produce γ-alumina from black aluminum dross is only considered within the defined system boundary. Since there is no consistent literature reported on the dross-to-water ratio, a ratio of 1:3 dross-to-water is considered in this water leaching process [21, 27]. In addition, only reactions involving Al_2O_3, AlP, and AlN compounds are being considered, given these compounds predominantly dictate the secondary aluminum dross composition [23, 39].

Since the electricity generation sources in Malaysia are mainly dominated by natural gas mix, the energy flow and its effects on the overall environmental footprint to produce γ-alumina are considered. In our study, 1,000 kg production of γ-alumina has been considered as the basis, and the energy consumption is scaled up based on the laboratory-scale study. The sensitivity analysis considers the following two scenarios: (i) leaching efficiency of 30%, 50%, and 70% for both HCl and H_2SO_4 leaching processes and (ii) comparison of the environmental impact between the non-renewable-based electricity and renewable-derived electricity (i.e. biogas and hydropower).

5. Conclusions

Life cycle assessment to produce γ-alumina from secondary aluminum dross via two acid leaching processes, namely HCl and H_2SO_4 acids, is carried out. In this assessment, electricity (natural gas) remains a major contributor in four out of five environmental impact categories assessed, namely global warming, acidification, human toxicity, and marine aquatic ecotoxicity categories. Sensitivity analysis with regard to leaching efficiency variations (30%, 50%, and 70%) has shown good improvements in acidification (36.4% improvement) and eutrophication (43.5% improvement) categories, while the remaining environmental categories have insignificant improvement. Nevertheless, transitioning to renewable energy sources, such as hydropower, contributes to a major improvement in most categories assessed due to the significant reduction in CO_2 emission. Nevertheless, utilization of biogas energy source gave retrograde results instead due to the high concentration of hydrogen sulfide that contributes to high acidification potential. Meanwhile, excess nitrate and benzene emissions are responsible for high eutrophication and human toxicity potential impacts. Going forward, substitution with less intensive processes with low energy requirements to produce γ-alumina from secondary aluminum dross continues to be a daunting challenge, and switching from a fossil-based to renewable energy source should be further considered to reduce the greenhouse gas emissions.

Acknowledgments

The authors would like to acknowledge the Ministry of Higher Education Malaysia for awarding the Higher Institution Centre of Excellence award to the Centre for Biofuel and Biochemical Research, Universiti Teknologi PETRONAS (Grant no. 015MA0-052 and no. 015MA0-104) as well as Mr. Yau Wen Zhen and Mr. Muhammad Naqiyuddin Azhari for data gathering.

References

[1]   R. Naderi, M. Fedel, F. Deflorian, M. Poelman, and M. Olivier, Synergistic effect of clay nanoparticles and cerium component on the corrosion behavior of eco-friendly silane sol–gel layer applied on pure aluminum. *Surf. Coat. Technol.* **224**, 93–100 (2013).

[2]   A. Algahtani and E. R. Mahmoud, Erosion and corrosion resistance of plasma electrolytic oxidized 6082 aluminum alloy surface at low and high temperatures. *J. Mater. Res. Technol.* **8**, 2699–2709 (2019).

[3]   T. Watari, K. Nansai, and K. Nakajima, Major metals demand, supply, and environmental impacts to 2100: A critical review. *Resour Conserv. Rec.* **164**, 105107 (2021).

[4]   U. Sultana, F. Gulshan, M. Gafur, and A. Kurny, Kinetics of recovery of alumina from aluminium casting waste through fusion with sodium hydroxide. *Am. J. Mater. Sci.* **1**, 30–34 (2013).

[5]   A. Meshram, R. Jha, and S. Varghese, Towards recycling: Understanding the modern approach to recover waste aluminium dross. *Mater. Today* **46**, Part 3, 1487–1491 (2021).

[6]   A. Tripathy, S. Mahalik, C. Sarangi, B. Tripathy, K. Sanjay, and I. Bhattacharya, A pyro-hydrometallurgical process for the recovery of alumina from waste aluminium dross. *Miner. Eng.* **137**, 181–186 (2019).

[7]   M. Mahinroosta and A. Allahverdi, Hazardous aluminum dross characterization and recycling strategies: A critical review. *J. Environ. Manage.* **223**, 452–468 (2018).

[8]   E. David and J. Kopac, Hydrolysis of aluminum dross material to achieve zero hazardous waste. *J. Hazard. Mater.* **209**, 501–509 (2012).

[9] A. Meshram and K. K. Singh, Recovery of valuable products from hazardous aluminum dross: A review. *Resour Conserv. Recycl.* **130**, 95–108 (2018).

[10] E. Petavratzi and S. Wilson, Residues from aluminium dross recycling in cement. *Characterization Miner. Wastes Resour. Process. Technol.* 1–8 (2007). Report no. is WRT 177 / WR0115.

[11] E. Elsarrag, A. Elhoweris, and Y. Alhorr, The production of hydrogen as an alternative energy carrier from aluminium waste. *Energ. Sustain. Soc.* 7, 1–14 (2017).

[12] D. O. Nduka, O. Joshua, A. M. Ajao, B. F. Ogunbayo, and K. E. Ogundipe, Influence of secondary aluminum dross (SAD) on compressive strength and water absorption capacity properties of sandcrete block. *Cogent Eng.* 6, 1608687 (2019).

[13] K. Hu, D. Reed, T. J. Robshaw, R. M. Smith, and M. D. Ogden, Characterisation of aluminium black dross before and after stepwise salt-phase dissolution in non-aqueous solvents. *J. Hazard. Mater.* **401**, 123351 (2021).

[14] K. Y. Paranjpe, Alpha, beta and gamma alumina as catalyst. *Pharm. Innov. J.* 6, 236–238 (2017).

[15] F. A. Perras, J. D. Padmos, R. L. Johnson, L.-L. Wang, T. J. Schwartz, T. Kobayashi, J. H. Horton, J. A. Dumesic, B. H. Shanks, and D. D. Johnson, Characterizing substrate–surface interactions on alumina-supported metal catalysts by dynamic nuclear polarization-enhanced double-resonance NMR spectroscopy. *J. Am. Chem. Soc.* **139**, 2702–2709 (2017).

[16] M. Mahinroosta, A. Allahverdi, P. Dong, and N. Bassim, Green template-free synthesis and characterization of mesoporous alumina as a high value-added product in aluminum black dross recycling strategy. *J. Alloys Compd.* **792**, 161–169 (2019).

[17] P. Tsakiridis, P. Oustadakis, and S. Agatzini-Leonardou, Aluminium recovery during black dross hydrothermal treatment. *J. Environ. Chem. Eng.* 1, 23–32 (2013).

[18] X. Zhu, Q. Jin, and Z. Ye, Life cycle environmental and economic assessment of alumina recovery from secondary aluminum dross in China. *J. Clean. Prod.* **277**, 123291 (2020).

[19] A. Meshram, A. Jain, M. D. Rao, and K. K. Singh, From industrial waste to valuable products: Preparation of hydrogen gas and alumina from aluminium dross. *J. Mater. Cycles Waste Manag.* **21**, 984–993 (2019).

[20] V. Kevorkijan, The quality of aluminum dross particles and cost-effective reinforcement for structural aluminum-based composites. *Compos. Sci. Technol.* **59**, 1745–1751 (1999).

[21] H. Feng, G. Zhang, Q. Yang, L. Xun, S. Zhen, and D. Liu, The investigation of optimizing leaching efficiency of al in secondary aluminum dross via pretreatment operations. *Processes* **8**, 1269 (2020).

[22] P. Li, M. Guo, M. Zhang, L. Teng, and S. Seetharaman, Leaching process investigation of secondary aluminum dross: The effect of CO_2 on leaching process of salt cake from aluminum remelting process. *Metall. Mater. Trans. B* **43**, 1220–1230 (2012).

[23] B. Das, B. Dash, B. Tripathy, I. Bhattacharya, and S. Das, Production of η-alumina from waste aluminium dross. *Miner. Eng.* **20**, 252–258 (2007).

[24] M. Mahinroosta and A. Allahverdi, Production of nanostructured γ-alumina from aluminum foundry tailing for catalytic applications. *Int. Nano Lett.* **8**, 255–261 (2018).

[25] Y. F. Adans, A. R. Martins, R. E. Coelho, C. F. d. Virgens, A. D. Ballarini, and L. S. Carvalho, A simple way to produce γ-alumina from aluminum cans by precipitation reactions. *Mater. Res.* **19**, 977–982 (2016).

[26] R. Matjie, J. Bunt, and J. Van Heerden, Extraction of alumina from coal fly ash generated from a selected low rank bituminous South African coal. *Miner. Eng.* **18**, 299–310 (2005).

[27] M. H. Abd Aziz, M. H. D. Othman, N. A. Hashim, M. A. Rahman, J. Jaafar, S. K. Hubadillah, and Z. S. Tai, Pretreated aluminium dross waste as a source of inexpensive alumina-spinel composite ceramic hollow fibre membrane for pretreatment of oily saline produced water. *Ceram. Int.* **45**, 2069–2078 (2019).

[28] E. Dekker, M. C. Zijp, M. E. van de Kamp, E. H. Temme, and R. van Zelm, A taste of the new ReCiPe for life cycle assessment: Consequences of the updated impact assessment method on food product LCAs. *Int. J. Life Cycle Assess.* **25**, 2315–2324 (2020).

[29] C. L. Yiin, S. Yusup, A. T. Quitain, Y. Uemura, M. Sasaki, and T. Kida, Life cycle assessment of oil palm empty fruit bunch delignification using natural malic acid-based low-transition-temperature mixtures: A gate-to-gate case study. *Clean Technol. Envir.* **20**, 1917–1928 (2018).

[30] T. Reyes, R. P. Gouvinhas, B. Laratte, and B. Chevalier, A method for choosing adapted life cycle assessment indicators as a driver of environmental learning: A French textile case study. *Artif. Intell. Eng. Des. Anal. Manuf.* **34**, 68–79 (2020).

[31] M. S. Reddy and D. Neeraja, Aluminum residue waste for possible utilisation as a material: A review. *Sadhana* **43**, 1–8 (2018).

[32] D. Brough and H. Jouhara, The aluminium industry: A review on state-of-the-art technologies, environmental impacts and possibilities for waste heat recovery. *Int. J. Thermofluids* **1–2**, 100007 (2020).

[33] F. Samara, T. Ghalayini, N. Abu Farha, and S. Kanan, The photocatalytic degradation of 2, 3, 7, 8-tetrachlorodibenzo-p-dioxin in the presence of silver–titanium based catalysts. *Catalysts* **10**, 957 (2020).

[34] N. Thanakunpaisit, N. Jantarachat, and U. Onthong, Removal of hydrogen sulfide from biogas using laterite materials as an adsorbent. *Energy Procedia* **138**, 1134–1139 (2017).

[35] H. Sawalha, M. Maghalseh, J. Qutaina, K. Junaidi, and E. R. Rene, Removal of hydrogen sulfide from biogas using activated carbon synthesized from different locally available biomass wastes-a case study from Palestine. *Bioengineered* **11**, 607–618 (2020).

[36] I. M. Andersen, T. J. Williamson, M. J. González, and M. J. Vanni, Nitrate, ammonium, and phosphorus drive seasonal nutrient limitation of chlorophytes, cyanobacteria, and diatoms in a hyper-eutrophic reservoir. *Limnol. Oceanogr.* **65**, 962–978 (2020).

[37] Y. Li, C. P. Alaimo, M. Kim, N. Y. Kado, J. Peppers, J. Xue, C. Wan, P. G. Green, R. Zhang, and B. M. Jenkins, Composition and toxicity of biogas produced from different feedstocks in California. *Environ. Sci. Technol.* **53**, 11569–11579 (2019).

[38] H. Pobeheim, B. Munk, H. Lindorfer, and G. M. Guebitz, Impact of nickel and cobalt on biogas production and process stability during semi-continuous anaerobic fermentation of a model substrate for maize silage. *Water Res.* **45**, 781–787 (2011).

[39] M. Türk, M. Altıner, S. Top, S. Karaca, and C. Bouchekrit, Production of alpha-alumina from black aluminum dross using NaOH leaching followed by calcination. *JOM* **72**, 3358–3366 (2020).

Chapter 8

Environmental and Cost Assessment of Spent Methanol via Life Cycle Assessment

Alvin W. L. EE[*,‡] and Valerio ISONI[†,§]

**Energy Studies Institute, National University of Singapore,
29 Heng Mui Keng Terrace, #0-01, Singapore 119620*

*†Institute of Chemical and Engineering Sciences, A*STAR,
1 Pesek Road, Jurong Island, Singapore 627833*

‡esiaewl@nus.edu.sg

§isoniva@ices.a-star.edu.sg

Spent solvents are one of the largest waste streams generated by pharmaceutical and chemical industries. In the effort to reduce waste and at the same time promote the recovery of resources, different waste treatment options of spent solvents are explored. A "gate-to-gate" Life Cycle Assessment approach is employed to study the environmental and cost assessment of the treatment of three different compositions of spent methanol. The treatment methods considered are as follows: (i) recovery through distillation, (ii) steam generation through co-generation (combustion), and (iii) disposal via specialized hazardous waste treatment facility. As cost factors can be a key consideration for decision-making, sensitivity analysis was carried out to determine how the various technologies will be influenced by price fluctuation. This work demonstrated that the recycling of spent methanol has proven to have the least impacts and cost, while utilizing spent solvent as a fuel has shown the benefits of generating surplus energy. The disposal of spent methanol via a hazardous waste treatment facility resulted in the highest environmental impact and cost involved.

1. Introduction

Spent solvents are reported to be the largest waste streams generated by pharmaceutical and chemical industries [1]. According to Raymond *et al.* [2], around 80–90% wt% solvents are used in the material input for the manufacture of active pharmaceutical ingredients (APIs). This makes spent solvent recovery a favorable option if environmental loads can be reduced. Solvent recovery can be carried out via an array of different physical separation methods depending on the solvent characteristics. It was cautioned by Luis *et al.* [3] that as an energy-intensive process, solvent recovery may result in unexpected negative impacts to the environment. Therefore, it is important that a Life Cycle Assessment (LCA) approach is employed to fully understand the potential environmental impacts of waste treatment options. In other near similar waste solvent treatment studies, Schott and Cánovas [4] highlighted that the environmental impacts of LCA may differ according to system boundary delimitations and assumptions.

In this chapter, the environmental impacts and potential costs of three scenarios of 25, 50, and 75 wt% spent solvent methanol from a waste stream will be investigated: (i) solvent recovery, (ii) energy recovery via co-gen, and (iii) disposal via specialized hazardous waste treatment facility, respectively.

1.1. *Solvent recovery*

To date, distillation remains the most common technology for solvent recovery [5], which is the focus of this work as the first option. In this process, heat input is required, which is typically supplied via a saturated steam or a thermal fluid such as hot oil. The recovered solvent which meets the purity requirement will reduce the need for purchasing new solvents, reducing the material cost for production, but may impact the cost requirement associated with the recovery process such as utilities. An overview of the operating parameters affecting the potential environmental and cost impacts of solvent recovery systems is tabulated in Table 1.

Table 1.   Summary of parameters of solvent distillation.

Generation of heating utilities	Composition of desired solvent in spent solvent mixture	If the desired solvent exists as low concentration, more energy will be required to heat non-desired component(s)
	Physical properties of desired solvent	If the desired solvent has a high latent heat requirement, more energy will be required to bring the spent solvent to boil for separation
	Operating pressure	Lower operating pressure will reduce the amount of heat energy required for separation
	Complexity of system	A distillation train may be required if a single distillation process is not effective in solvent recovery
Generation of cooling unities	Type of cooling utility used	Depending on the cooling utility used for the condensation process, the potential environmental impact changes, for example, using ambient water versus using chilled water
	Physical properties of desired solvent	If the desired solvent has a high latent heat requirement, more cooling duty is required to condense the solvent vapor to liquid
Electricity consumption	Condensation method adopted	Air-cooled condenser may require electricity to drive the motor for a fan system
	Equipment duty	Equipment used in the solvent recovery process such as pumps consumes electricity
(De)pressurizing of distillation column	Methods adopted to (de)pressurize distillation column	Potential environmental impact associated with methods of (de)pressurizing the distillation column varies
Solvent top-up	Efficiency of distillation	A more efficient distillation system will minimize desired solvent in waste stream, reducing the amount of top-up required
Entrainer/ solvent	Properties of component in mixture	Additional solvent (entrainer) may be required for azeotropic mixture
Operating cost	Solvent for top-up	Efficiency of recovery process
	Entrainer/additional solvent	Depending on the complexity of recovery system such as chemical properties, ease of separation, etc.
	Energy use	Depending on the complexity of system such as boiling point of component in spent solvent mixture, purity of product required, etc.

1.2. *Solvent-to-energy* (*steam co-generation*)

High calorific spent solvents may be potentially used for on-site steam generation [6, 7], but this approach may pose a challenge for the constant supply of steam if the spent solvent is generated in batches. In order to ensure a continuous supply of spent solvent for steam generation, companies can collaborate with others within their proximity and buy their spent solvents on a regular basis. An overview of the operating parameters associated with on-site steam generation using spent solvent is tabulated in Table 2.

Table 2.   Summary of parameters associated with in-situ steam generation.

Fresh solvent required	Amount of fresh solvent required by main production process	Different solvents will result in different potential environmental impacts associated with its manufacturing. The impacts are amplified by the amount required
Emissions	Substances present	The type of substances present may result in different off-gas properties which can affect the potential environmental impacts
	Efficiency of boiler	Ineffective combustion may produce monoxides
	Efficiency of scrubber	Ineffective scrubbers may result in higher release of acidic gases
Electricity consumption	Equipment duty	Equipment used in the solvent recovery process such as pumps requires electrical input
Operating cost	Amount of fresh solvent required	Affected by market price
	Diesel	In the presence of chlorinated solvents, higher combustion temperatures are required. Diesel may be required if the spent solvent flame temperature does not meet the minimum combustion temperature set by local authorities
	Energy	Number of equipment required by in-situ steam generation which requires energy and other related input

1.3. *External disposal via incineration (without energy recovery)*

The third option is the disposal of spent solvent. In this last option, waste chemicals or solvents are typically sent to waste treatment facilities [8]. In addition, this is also particularly preferred by firms with small waste volumes or inconsistent spent solvents, as it eliminates the need for in-house waste treatment which may be costly due to large capital investment required.

2. LCA System

The functional unit of 1,000 tonne/year (tpy) spent solvent methanol is used as the comparative basis. The "gate-to-gate" LCA system is illustrated in Figure 1. The potential environmental impacts of 25, 50, and 75 wt% spent methanol organic stream from a typical pharmaceutical manufacturing industry will be investigated. The following waste scenarios are compared:

- Scenario 1 (S-1): Solvent recovery of 25, 50, and 75 wt% spent methanol
- Scenario 2 (S-2): Steam generation from the combustion of 25, 50, and 75 wt% spent methanol
- Scenario 3 (S-3): Disposal via incineration (without energy recovery) of 25, 50, and 75 wt% spent methanol.

The life cycle inventory data were obtained from the literature and scientifically accepted database (Ecoinvent) [9]; additional information, such as the operations involved and associated energy requirements for solvent recovery systems, was obtained using simulation carried out in *ASPEN HYSYS*. In order to perform a simplified economic analysis, the following were considered: cost of methanol ($551.79/tonne), cost of steam ($14.78/tonne), and waste disposal fee for spent methanol ($1.00/liter).

The LCA investigation for the waste management system (Figure 1) is carried out according to the following conditions:

- In each option, the waste material flow is considered to be running in a continuous operation system

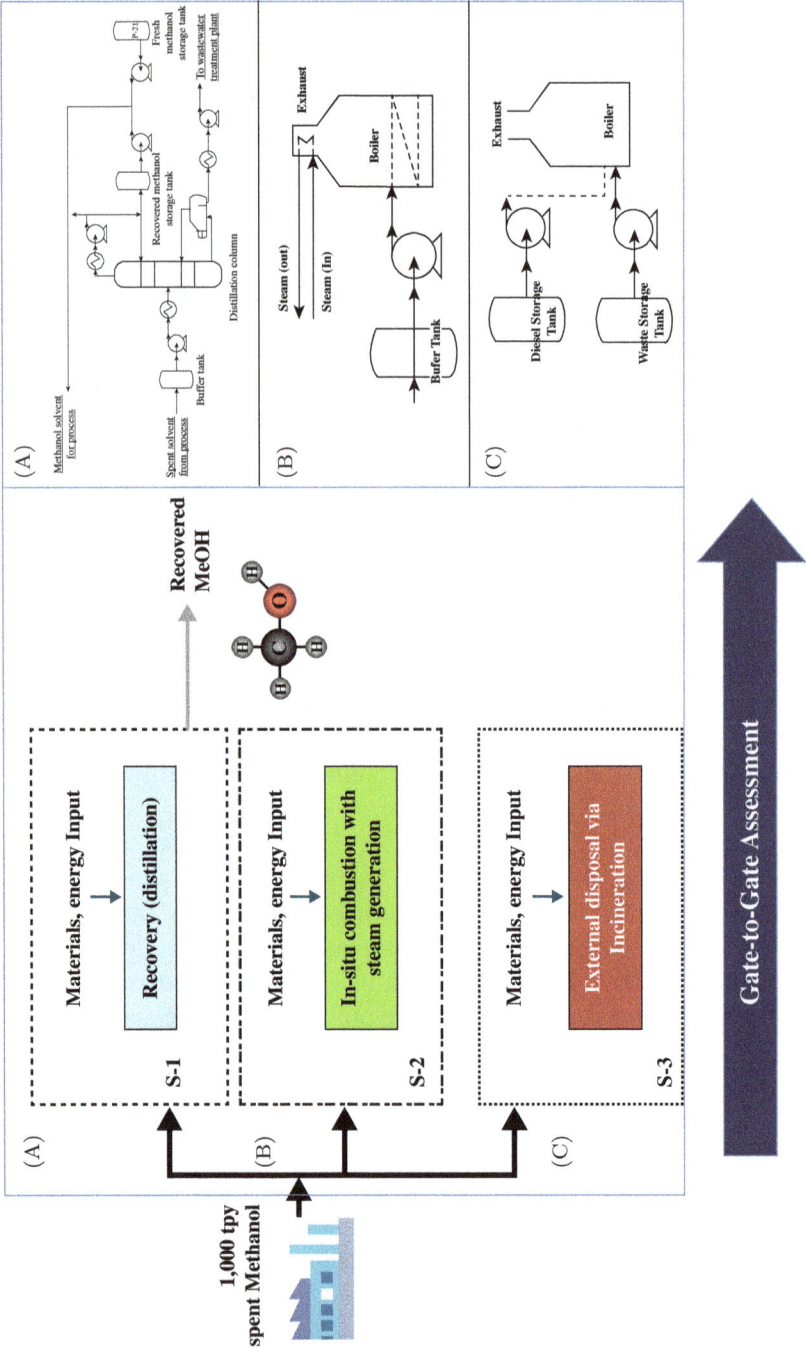

Figure 1.   "Gate-to-gate" LCA system with scenarios of (A) distillation, (B) *in-situ* combustion, and (C) disposal.

- All modes of transportation are not taken into account
- Mass co-allocation is applied to measure the operating parameters and emissions of the spent solvent
- It is assumed that in S-3, the high calorific value of spent methanol (22.7 MJ/kg) supplies heat to support the incineration process
- The potential operating costs involved in all three waste management options are considered for the unit price of (additional) methanol required for S1, steam (utilities), and waste disposal fees.

3. Results and Discussions

The following environmental impacts are generated with the CML 2001 life cycle impact method:

- Global Warming Potential or GWP (in kg CO_2-eq)
- Acidification (in kg SO_2-eq)
- Eutrophication (in kg phosphate-eq)
- Human Toxicity (in kg DCB-eq)
- Photochemical Ozone Creation Potential or POCP (in kg ethene-eq)
- Energy (used and surplus) in GJ

Included in the results are the total costs of methanol, steam, and waste disposal fees where necessary. The results (excluding energy used and surplus) for 25 wt%, 50 wt%, and 75wt% aqueous spent methanol comparing all three scenarios are presented in Figure 2(a)–(c). Energy spent for each scenario is depicted in Figure 3.

For S-1, the GWP results for the recovery of 25 wt%, 50 wt%, and 75 wt% spent methanol are 99, 125, and 151 tonne CO_2-eq/year, respectively. The generation of steam for the solvent recovery process is the main contributor of these impacts. It can also be observed that the environmental impacts of scenarios 2 and 3 are comparable, except for costs. The total annual costs for the external disposal of 1,000 tonne/year spent methanol are \$1,194,734/year (for 25 wt%), \$1,395,349/year (for 50 wt%), and \$1,604,145/year (for 75 wt%).

The corresponding annual costs for distillation are estimated to be \$8,478 (for 25 wt%), \$10,667 (for 50 wt%), and \$12,857 (for 75 wt%) per year. The main contributor of these costs is the purchase of steam to facilitate distillation (ca. 99%). It should be noted, however, that the costs do not take into account process operations, maintenance, or any ancillaries.

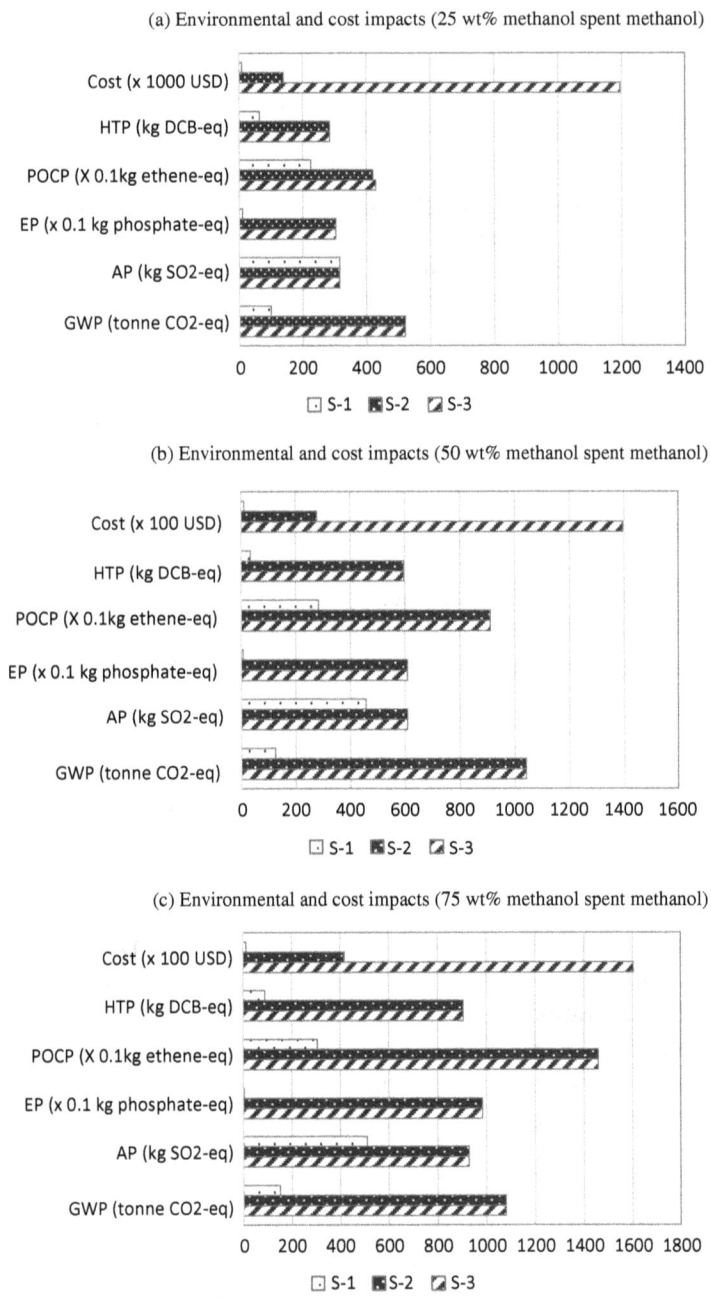

(a) Environmental and cost impacts (25 wt% methanol spent methanol)

(b) Environmental and cost impacts (50 wt% methanol spent methanol)

(c) Environmental and cost impacts (75 wt% methanol spent methanol)

Figure 2.   (a–c) Environmental and cost impact results for scenarios 1, 2, and 3 for (a) 25 wt%, (b) 50 wt%, and (c) 75 wt% spent methanol, respectively.

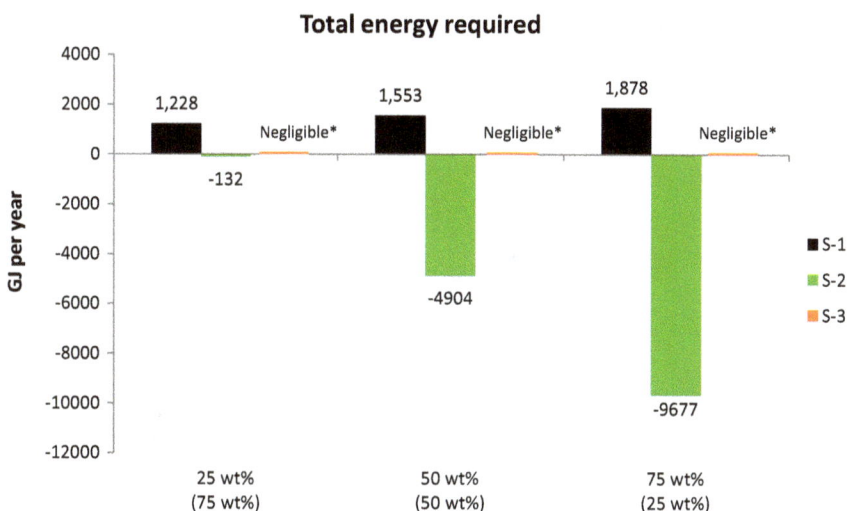

Figure 3.   Energy results for scenarios 1, 2, and 3.

Note: * Self-supply of heat is assumed to support the incineration process.

As anticipated, Figure 3 indicates that process energy demands are highest for scenario 1 (distillation of spent solvent). The amount of energy required to recover 1,000 tonne/year of 25 wt%, 50 wt%, and 75 wt% spent methanol is 1,228 GJ/year, 1,553 GJ/year, and 1878 GJ/year, respectively. Despite this, the environmental burdens associated with solvent recovery can be observed (Figure 2). S-2 displayed an advantage over S-1 in terms of energy recovery. The cumulative amount of energy gained from the on-site combustion of 25 wt%, 50 wt%, and 75 wt% spent methanol is 132 GJ/year, 4,904 GJ/year, and 9,677 GJ/year, respectively.

3.1. *Sensitivity analysis*

The potential cost fluctuation associated with material, disposal, and utility use may have a strong influence on decision-making for waste treatment options. Therefore, sensitivity analysis is carried out to determine the degree of cost result fluctuations.

3.1.1. *Cost of methanol*

The cost of methanol was adjusted from $552/tonne to the historical low and high values to determine variations of cost impacts on the three waste

Total cost results due to variations in price of methanol

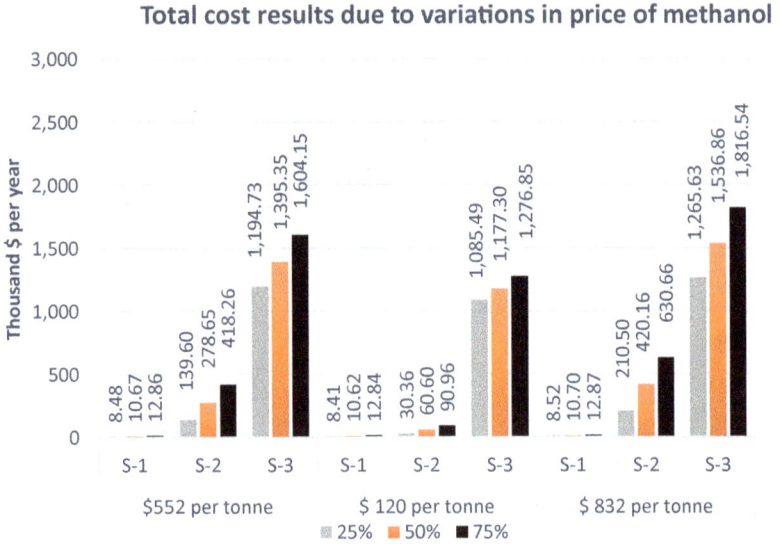

Figure 4.   Variations of overall cost results due to changes in methanol prices.

management scenarios. As illustrated in Figure 4, S-2 was observed to be the most sensitive to cost variations, as compared to the other two scenarios. The percentage changes for the overall cost results fluctuate from −78% to 51%. S-3 is less sensitive to the varying degree of methanol prices – the change in the overall cost is 6% to 20%. Solvent price variations contributed to only <1% of the total costs to S-1.

3.1.2. *Cost of utility (Steam)*

In the LCA model, steam is assumed to be produced by the combustion of natural gas; hence, the cost of steam is expected to be influenced by the price of natural gas. The unit price of steam was changed from $14.78 to $7.04 and $47.51. As steam is required to facilitate the solvent recovery process, the change in cost will affect S-1. The resultant total cost implications for S-1 are reported in Figure 5. The results show that the overall costs varied significantly from −50% to a 220% increment in cost. Nonetheless, despite the increase, scenario 1 still exhibits the lowest annual overall cost.

Total cost results due to variations in price of steam

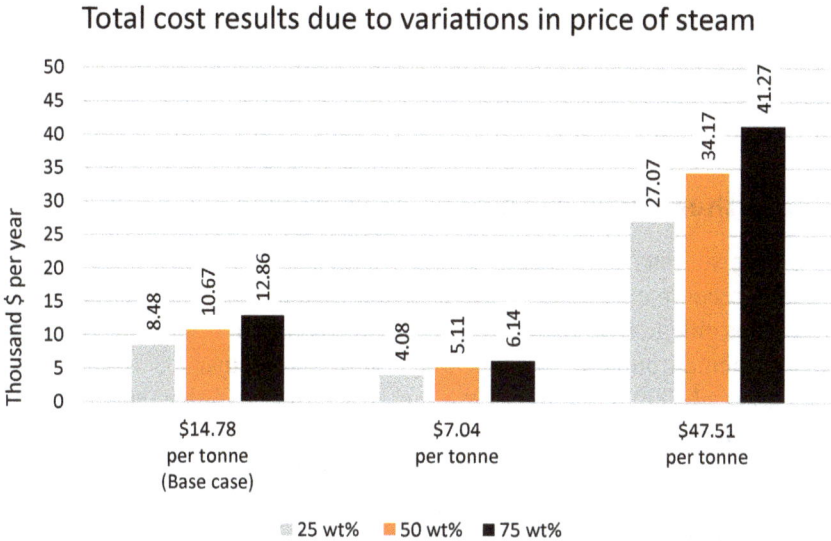

Figure 5. Variations of overall cost results due to changes in steam prices (scenario 1).

Total cost results due to variations in disposal fee

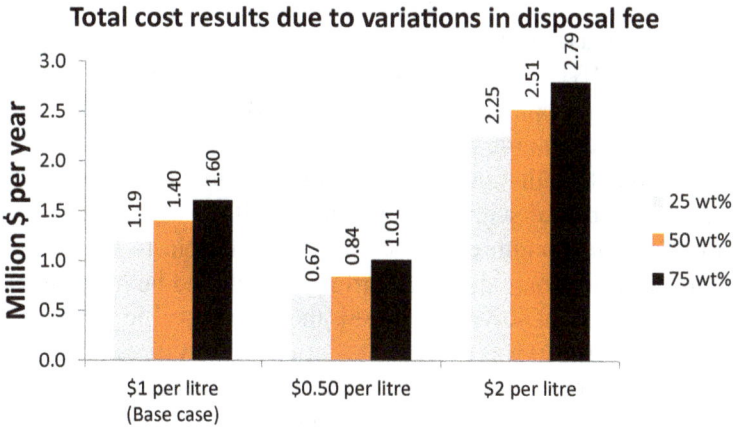

Figure 6. Variations of overall cost results due to changes in disposal fees (scenario 3).

3.1.3. *Disposal fee*

Disposal fee is expected to have a great influence on the total costs of S-3, and therefore the unit cost of $1 per liter was changed to $0.50 and $2. The resultant total cost implications for S-3 are reported in Figure 6.

Despite relatively small variations in disposal fees, the overall changes for S-3 are significant. The overall costs varied significantly from a reduction of −37% to an 88% increment in cost.

4. Further Discussions

In a near similar LCA study by Luis *et al.* [10] on waste solvents, the authors recommended that chemical compounds should be recovered where possible to help promote resource recovery and minimize potential environmental burdens generated by process industries. In addition, Capello *et al.* [11] recommended the use of energy-efficient and effective distillation recovery systems for spent solvent recovery as the preferred option over other waste management methods. This in turn reduces the life cycle impacts of fresh solvent production [12]. Our work demonstrates that apart from environmental performance, costs play an important role in selecting waste management options for spent solvents. In many such cases, priorities between environmental performance and cost performance have to be managed. As a noteworthy example, a combined LCA and cost assessment was carried out by Panepinto *et al.* [13] to examined two waste treatment options. The results of their work concluded that the MBT (mechanical biological treatment) option was preferable based on environmental performance, while pyro-gasification was favored from an economic point of view.

Spent solvents with high calorific values and high flame temperatures are likely to be mixed with other materials with lower calorific values/ adiabatic flame temperatures to facilitate incineration. In such cases, the emissions generated may differ accordingly. It should be highlighted that in the option of spent solvent disposal, the high calorific value of spent methanol (22.7 MJ/kg) was assumed to supply heat to support the disposal facility. Incineration emissions and energy requirements are likely to be dissimilar for other case studies (e.g. Seyler *et al.* [14] and Santoleri *et al.* [15]) due to different LCA modeling approaches, stages involved in system boundaries, types of waste compositions, and treatment options. For instance, Hofstetter *et al.* [16] evaluated the environmentally preferable treatment options considering incineration and distillation of waste solvent mixture containing toluene. In yet another case, Seyler *et al.* [17] evaluated the use of waste solvent as fuel substitute in the cement industry. In a more recent case study, conceptual model-based optimization processes were performed for the design of two waste solvent treatment technologies: (i) hybrid process distillation assisted by pervaporation and (ii) distillation/incineration [18].

5. Concluding Remarks

In recent years, apart from pollution reduction, resource conservation has become increasingly important, especially for the manufacturing of fossil-based chemicals. The objective of solvent recovery is very much aligned with this scheme. From the results, scenario 1 (solvent recovery) exhibits the overall benefits of both cost and environmental impact reductions. For GWP alone, reductions of 81%, 88%, and 86% are observed for 25 wt%, 50 wt%, and 75 wt% spent methanol, respectively, as compared to the GWP impacts of scenarios 2 and 3. It was also noted that scenario 1 resulted in more energy requirements — from 1,228 GJ/year (for the recovery of 25 wt% spent methanol) to 1,878 GJ/year (75 wt% methanol). This also implies that the costs of steam will greatly influence the overall cost results.

In the case where the generation of surplus energy is considered an advantage, scenario 2 becomes a preferred option. Up to 9,677 GJ/year of net energy can be gained from the on-site combustion of 75 wt% spent methanol. In terms of overall environmental impacts, scenarios 2 and 3 both exhibit the highest results, with scenario 3 exhibiting the highest costs. Apart from the environmental emissions caused by a hazardous waste treatment facility or incineration, the disposal fee charged by an external contractor contributes significantly to the high cost results.

While this chapter presents options for the treatment of spent methanol solvents, it is well acknowledged that the results may differ if different solvents or compositions are used. However, the study has shown that the adoption of a life cycle assessment tool provides the platform for multi-impact evaluation and allows businesses to make decisions from both the environmental and cost point of view.

References

[1] C. Jiménez-González, P. Poechlauer, Q. B. Broxterman, B. S. Yang, D. Am Ende, D. J. Baird, *et al.*, Key green engineering research areas for sustainable manufacturing: A perspective from pharmaceutical and fine chemicals manufacturers. *Org. Process Res. Dev.* **15**(4), 900–911 (2011).

[2] C. Seyler, T. B. Hofstetter, and K. Hungerbühler, Life cycle inventory for thermal treatment of waste solvent from chemical industry: A multi-input allocation model. *J. Clean. Prod.* **13**(13–14), 1211–1224 (2005).

[3] P. Luis, A. Amelio, S. Vreysen, V. Calabro, and B. Van der Bruggen, Simulation and environmental evaluation of process design: Distillation vs. hybrid distillation–pervaporation for methanol/tetrahydrofuran separation. *App. Energ.* **113**, 565–575 (2014).

[4] A.B.S Schott and A. Cánovas, Current practice, challenges and potential methodological improvements in environmental evaluations of food waste prevention–A discussion paper. *Res. Conserv. Rec.* **101**, 132–142 (2015).

[5] G. Van der Vorst, P. Swart, W. Aelterman, A. Van Brecht, E. Graauwmans, E., H. Van Langenhove, *et al.*, Resource consumption of pharmaceutical waste solvent valorization alternatives, *Res Conserv. Rec.* **54**(12), 1386–1392 (2010).

[6] P. N. Pressley, J. W. Levis, A. Damgaard, M. A. Barlaz, and J. F. DeCarolis, Analysis of material recovery facilities for use in life-cycle assessment. *Waste Manage.* **35**, 307–317 (2015).

[7] V. Ganapathy, *Industrial Boilers and Heat Recovery Steam Generators: Design, Applications, and Calculations* (CRC Press, New York, 2002).

[8] A. Amelio, G. Genduso, S. Vreysen, P. Luis, and B. Van der Bruggen, Guidelines based on life cycle assessment for solvent selection during the process design and evaluation of treatment alternatives. *Green Chem.* **16**(6), 3045–3063 (2014).

[9] G. Wernet, C. Bauer, B. Steubing, J. Reinhard, E. Moreno-Ruiz, E. and B. Weidema, The ecoinvent database version 3 (part I): Overview and methodology. *Int. J. Life Cycle Assess.* **21**(9), 1218–1230 (2016).

[10] P. Luis, A. Amelio, S. Vreysen, V. Calabro, and B. Van der Bruggen, Life cycle assessment of alternatives for waste-solvent valorization: Batch and continuous distillation vs incineration. *Int. J. Life Cycle Assess.* **18**(5), 1048–1061 (2013).

[11] C. Capello, S. Hellweg, B. Badertscher, H. Betschart, and K. Hungerbühler, Environmental assessment of waste-solvent treatment options. *J. Ind. Ecol.* **11**(4), 26–38 (2007).

[12] M. J. Raymond, C. S. Slater, and M. J. Savelski, LCA approach to the analysis of solvent waste issues in the pharmaceutical industry. *Green Chem.* **12**(10), 1826–1834 (2010).

[13] D. Panepinto, G.A. Blengini, and G. Genon, Economic and environmental comparison between two scenarios of waste management: MBT vs thermal treatment. *Res. Conserv. Rec.* **97**, 16–23 (2015).

[14] C. Seyler, T. B. Hofstetter, and K. Hungerbühler, Life cycle inventory for thermal treatment of waste solvent from chemical industry: A multi-input allocation model. *J. Clean. Prod.* **13**(13–14), 1211–1224 (2005).

[15] J. J. Santoleri, J. Reynolds, and L. Theodore, *Introduction to Hazardous Waste Incineration* (John Wiley & Sons, New York, 2000).

[16] T.B. Hofstetter, C. Capello, K. Hungerbühler, Environmentally preferable treatment options for industrial waste solvent management: A case study of a toluene containing waste solvent, *Process Safety Environ. Protect.* **81**, 189–202 (2003).

[17] C. Seyler, S. Hellweg, M. Monteil, and K. Hungerbühler, Life cycle inventory for use of waste solvent as fuel substitute in the cement industry — A multi-input allocation model. *Int. J. Life Cycle Assess.* **10**, 120–130 (2005).

[18] R. Meyer, D. A. F. Paredes, M. Fuentes, A. Amelio, B. Morero, P. Luis, *et al.*, Conceptual model-based optimization and environmental evaluation of waste solvent technologies: Distillation/incineration versus distillation/pervaporation. *Separat. Purific. Technol.*, **158**, 238–249 (2016).

Chapter 9

Integration of Life Cycle Assessment and Life Cycle Costing Methodology

Michal BIERNACKI

*Wroclaw University of Economics and Business,
ul. Komandorska 118/120, 53–345 Poland*

michal.biernacki@ue.wroc.pl

Environmental life cycle costing (ELCC) presents itself as a relatively new addition to the set of cost calculation models employed in the context of environmental management, sustainable growth policy, or corporate social responsibility. Through wide dissemination and popularization, it stands a good chance of becoming a universal system, a missing link integrating the varied aspects of environmental management as well as financial and bookkeeping duties. Such an effect can be obtained by a proper representation of all the product-related factors and properties that affect the environment, through its interaction with life cycle assessment (LCA). The model under study proves effective in the evaluation of already employed technologies, environmental chokepoints, planned technological improvements, and product/process modernizations or wind-downs, particularly in the context of technological effectiveness, social and environmental impact, and financial effects. As such, the model may prove particularly useful for production companies, regardless of their profile. The whole environmental life cycle costing (WELCC) formula postulated herein may prove not only valuable but also crucial due to its potential to integrate all the environmental aspects of the studied products or processes. Flexibility is another important argument for the implementation of environmental life cycle costing (ELCC), as the model may easily be adjusted to specific needs, offering substantial decision-making support at both the operating and the strategic level, based on a detailed database of information on any costs associated with the environmental impact of the studied products or processes.

1. Introduction

Practical realization of objectives and tasks related to the implementation of the concepts of sustainable development and corporate social responsibility with the purpose of ensuring continuous improvement of environmental effectiveness may be supported and supplemented by various mechanisms and instruments of environmental management accounting (EMA). The concept of sustainable development may easily be transposed from the macro scale to the level of individual companies through effective integration of the internal product policy that fulfills the criterion of environmental effectiveness of production at rationally determined energy and material outlays and at levels offering effective cost minimization for companies that choose to employ such an approach. Thus, corporate social responsibility may be defined as the company's accountability for the entirety of corporate activities associated with the practical implementation of this concept within the bounds of their environmental policies. The environmental policy, on the other hand, may be defined in terms of informed and responsible control over the company's effective impact upon the individual ecosystems and the natural environment as a whole. The effective environmental impact of products, in purely operating terms, should also be analyzed from the viewpoint of correlations between the production processes employed and the functional requirements of the natural environment. Economic entities are bound not only by restrictions and requirements of legal and economic character but also by those defined by marketing, business, political, and strategic dictates. Corporate management, in purely ecological aspects, should thus be analyzed in the following four fundamental dimensions that serve to support the effective market operation of economic entities:

- *The environmental dimension*: Reduction of natural resource consumption, defining the environmental effectiveness of company operations and limitation or deceleration of processes that generate high entropy.
- *The social dimension*: Ensuring wide acceptance and social legitimation for company activities.
- *The legal and political dimension*: Ensuring that company operations are conducted in accordance with the pending laws and regulations, particularly those related to environmental protection and observance of environmental standards.
- *The economic and market dimension*: Related to the increased cost of damage control, waste utilization, and implementation of environmental innovations [1].

According to the German BMU report [2, 3], the effective management of sustainable development will be most pronounced in the following areas:

- further reduction of energy and resource consumption (environmental demands),
- further increase of energy and resource effectiveness (environmental demands),
- effective management of resource and energy flows (environmental demands),
- increased transparency through improvements of environmental information systems in companies (environmental demands),
- further improvement of corporate social impact through the use of dedicated instruments (social demands),
- propagation and dissemination of social communication instruments to better meet the requirements of the stakeholders (social demands),
- increased use of process accounting and material flow accounting methods (economic demands),
- budgeting of environmental and social costs at the company level (economic demands),
- improved transparency of the increase in company effectiveness and value resulting from the use of proper sustainable development management instruments [2, 3].

Based on the above, it seems that companies should focus on improving the integration between the various demands of sustainable development described above to meet the practical requirements defined by the market competition, the formal regulations, and the social expectations. One of the ways to approach the task of collating information for effective sustainable development management is to combine the methods of life cycle analysis (LCA) and life cycle costing (LCC) — such integration is offered by the environmental life cycle costing (ELCC) instrument postulated herein.

2. Life Cycle Costing: Fundamental Assumptions

LCC is an example of phase III autonomous system, designed on a basis of two cost accounting systems kept in parallel, one oriented to meet the requirements of external reporting and the other designed to support the internal strategic decision-making processes. Incorporation of LCC at the company level offers a suitable fundament of information required to make rational decisions of both strategic and operating character. However, this approach is also described by an ostensive lack of evaluations related to the realization of policies of sustainable development at the company level. Various research efforts have been carried out to incorporate LCC modeling for different evaluative purposes, including material or resource selection, energy conservation programs, R&D incentives, and environmental policy planning [4–7].

However, most literature studies fail to provide a widely acceptable and cohesive definition of life cycle costing. The historical development of this concept, both in terms of its theoretical base and the economic practice, has stimulated a wealth of widely dispersed definitions and interpretations. The first normalized definition of the term was provided by Haworth [8]; in his approach, LCC represents a process involving the collection of information on any and all costs associated with each decision and manifested throughout the entire lifetime of specific processes [8]. Ansari and Bell [9], on the other hand, defined LCC as a cost account of all activities undertaken in relation to the product, from the initial idea up to the end of the product market lifetime. In addition, Ansari and Bell [9] argued that a product life cycle does not culminate in sales closure, as many companies are likely to continue their service and post-sales duties for products well past their decommission date. Woodward [10] defined

LCC of an asset as an account taking into consideration all the cost factors relating to the asset during its operational life, from the development/procurement stage to eventual disposal. The American Institute of Logistical Management presents an asset's life cycle cost as a sum of all costs related to the owner's control over the asset, including those associated with the asset's production or procurement, business use and operation, maintenance, staff training, and liquidation [11]. Baussabaine and Kirkham [12] strongly advised that the use of a whole life cycle costing (WLCC) should be implemented. Aside from cost items related directly to the product under evaluation, the WLCC method emphasizes the evaluation of less indirect costs, including cost of capital, cost of risk, and environmental costs, with major focus on the selection of adequate forecasting purposes. It must be noted that the term "product" (in the above contexts) should be interpreted as a reference to a class of products — rather than a single item — and that it may refer to both physical products and services.

As evidenced by the above, the need for a generally acceptable and objective model of LCC is quite apparent. In response to this demand, the most generalized and the most universal approach to LCC modeling may be postulated as follows:

$$TLCC = DPC + RTC + SC + PC + SMC + RMC + OC + PLC + PDC \tag{1}$$

where TLCC denotes the total product life cycle cost, DPC the design and planning costs, RTC the research and testing costs, SC the supply costs, PC the production (service rendering) costs, SMC the sales and marketing costs, RMC the repair and maintenance costs (as borne by the producer), OC the other costs generated in the market phase, PLC the product liquidation costs, and PDC the product discontinuation costs.

The costing items from Equation (1) may be further analyzed in detail. For example, the total cost of research and testing may further be segregated into the following cost item groups:

- software,
- testing,
- program management,
- conceptual research,
- postulated project changes,

- project engineering,
- technical data management,
- staff training.

The principal purpose of LCC is to provide information for analyses of costs incurred in relation to specific products in each phase of their life cycle, from the early ideation and design, through the operating phase, through product discontinuation phase (product death), to the phase of product utilization and disposal. By employing the above delimitations, producers are better equipped to devise alternatives for their products or to track the dynamics of market changes. The most obvious benefits of LCC include comprehensive presentation of a decision-making dilemma in a time-based approach, analyses of cost structure in each phase of a product life cycle, long-term forecasts of trends, and potential for introducing curative measures at any point in the practical realization of strategies.

The above benefits are directly contributable to the fact that LCC is — in a sense — a compilation of several distinct approaches to cost calculation, employed for specific contexts of each of the phases of the product life cycle. Thus, it may safely be postulated that the phases of R&D, market introduction, and growth shall typically employ the following methods:

- traditional cost accounting,
- target costing,
- activity-based costing,
- quality costing (cost of quality — COQ),
- logistics costing,
- kaizen costing (cost of continued improvement),
- time-driven activity-based costing.

On the other hand, the phases of product maturity, decline and liquidation shall be best described by the following:

- traditional cost accounting,
- activity based costing,
- value chain cost management accounting,
- quality costing,

- logistics costing,
- time-driven activity-based costing.

3. Environmental Costs: Fundamental Assumptions

In the wide area of cost concerns associated with environmental management, LCC-oriented environmental accounting tools have been suggested
[13–15]. Depending on the adopted definition and practical approach, the
environmental accounting may serve to supplement and support not only
the traditional budget accounting in a macroeconomic dimension but also
the internal processes of financial accounting and management accounting in individual entities. Taking into consideration the character and
specificity of environmental protection, a division may be postulated
involving two separate dimensions:

- the internal dimension
  o internal environmental costs,
- the external dimension
  o external environmental costs,
  o external social costs.

The first dimension represents those of the environmental costs
incurred by entities which directly influence their financial result. These
are typically reflected in company bookkeeping records. The second
dimension relates to costs incurred or borne by other physical persons or
the society in general as well as off-balance sheet costs borne by the
entity. External costs are typically included in macroeconomic and environmental accounting reports, but many entities are exempt from such
duties. The external environmental cost for an entity is a pecuniary
expression of total negative environmental effects related to waste and
pollution. These effects are generated externally (but are attributable to
product manufacturing processes) and — as such — should be reflected
as off-balance sheet items. In the process of determining the individual
cost groups, it may be useful to assume that fossil fuels can be replaced
by renewable energy sources; this type of cost is generated externally and
should be reflected as an off-balance sheet item; calculations may be
based on readily available databases, such as those published by SETAC
and Design 4 Sustainability. The following formula may be adopted for

the calculation of external environmental cost for each phase of a product life cycle:

$$EEC = ES \times WP \tag{2}$$

where EEC denotes the external environmental cost in a given phase of a product life cycle, ES the volume of waste production or air pollutant emission (e.g. SO_2), obtained from LCA technique calculations, and WP the conversion coefficient of waste production or air pollutant emission (e.g. 14.67 EUR/kg SO_x based on Eco Cost Database available at http://www.design-4-sustainability.com/ecocosts).

It was assumed that the external environmental cost represents this part of costs which arises from the entity's impact upon society in general without bearing any pending financial responsibilities for the entity in question (such obligation may arise in the future, imposed by legislative changes or through reformulation of corporate social responsibility strategies). As such, this part of cost has no immediate effect upon company results. These costs are generated externally and should be reported as off-balance sheet items or — in line with the prudency principle — as reserves to cover for any future financial obligations that may arise. Calculations of this part of costs over the course of a given product life cycle may be obtained from the following formula:

$$SC = P \times U \times O \tag{3}$$

where SC denotes the social cost, P the probability of a given cost arising, U the estimated volume of cost, calculated on the basis of average compensation offered by insurance companies in applicable conditions (e.g. 30,000 EUR for patients with diagnosed cancer), average land remediation prices, or other damage-related costs, and O an estimated size of population with potential claim rights.

Environmental costs generated internally should also deserve special managerial attention for the following reasons (among others):

- A good portion of such costs may be controlled or even eliminated through correct identification of technological attributes and through informed managerial decisions in various aspects, from operational changes, through pro-environmental investments, up to radical reformulation or redesign of processes or products.

- Environmental costs are often inventoried as off-balance sheet items.
- Environmental costs may be compensated by revenue from marginal production or from the sale of tradeable emission allowances.
- Improvement of managerial control over environmental costs may substantially increase the entity's ability to meet the adopted environmental criteria but also induce positive changes in public health as well as contribute in the entity's economic success through eco-marketing.
- Good understanding of environmental costs, process effectiveness, and product specificities offers potential to promote detailed calculation of costs and product prices, while at the same time providing fundaments for the design and improvement of environmentally friendly processes, products, and services.
- Placing emphasis on the environmental concerns through separation of environmental costs in the processual determination of product prices has the effect of increasing the entity's competitive market advantage and presents a substantial contribution in the process of building a positive public image.
- Implementation of environmental cost accounting procedures may serve to support the development and effective operation of environmental management systems, particularly those designed on the basis of ISO standards.
- Keeping track of benefits and costs associated with environmental aspects of company operation may serve to support the effective allocation of resources with the view of meeting the environmental objectives.

In many scenarios, the environmental costs may be drastically reduced or eliminated through incorporation of best practices, such as changes in product design, the use of resource substitutes at input, redesign of processes and improvement of operational and maintenance procedures (to name a few). In addition, this part of costs may sometimes be manifested only after the effective volume of emission or waste produced by the entity passes a certain referential level expressed in applicable standards. Thus, by introducing an effective reduction of the use of chemical substrates below a recommended threshold or by using substitutes for such resources, an entity stands a good chance of reaping sizeable benefits in terms of cost reduction, even as early as in the initial phase of product design.

Entities with effective implementations of formal environmental management systems may institutionalize their environmental accounting tasks, as this type of instrument offers support for decisions made with respect to such systems. In addition, environmental accounting may play an important role in the following processes and tasks:

- cost management,
- management of activity-based tasks,
- quality management (including environmental quality management),
- process analyses,
- supply chain analyses,
- pro-environmental projects,
- design and evaluation of product life cycles,
- product life cycle costs.

4. Environmental Life Cycle Costing: Implementation of Life Cycle Assessment and Life Cycle Costing Combined

ELCC — in its fundamental assumptions — represents an attempt to evaluate the economic dimension of production from the viewpoint of its environmental impact expressed by the emergence of environmental costs. It also satisfies the requirements of pro-environmental production within the conceptual framework of sustainable development. Therefore, it bears the attributes of a comprehensive method and is designed to provide a consistent approach to products, much like the planning phase employed within the traditional framework of LCC. Schau *et al.* [16] presented a noteworthy stepwise procedure considering LCA as the main method, followed by LCC concepts as the second part of CSR (Figure 1).

LCC and LCA can be applied to measure cost and environmental performance of activity of all types of enterprises while assimilating information from two different disciplines (nature and social science). Hunkeler *et al.* [17] proposed that a general structure of ELCC is based on the physical life cycle of a product. This method needs separate assessment of the following five distinct life time stages, detailed and elaborated as required: research and development, production, use and maintenance, and disposal/recycling management, as described in Figure 2.

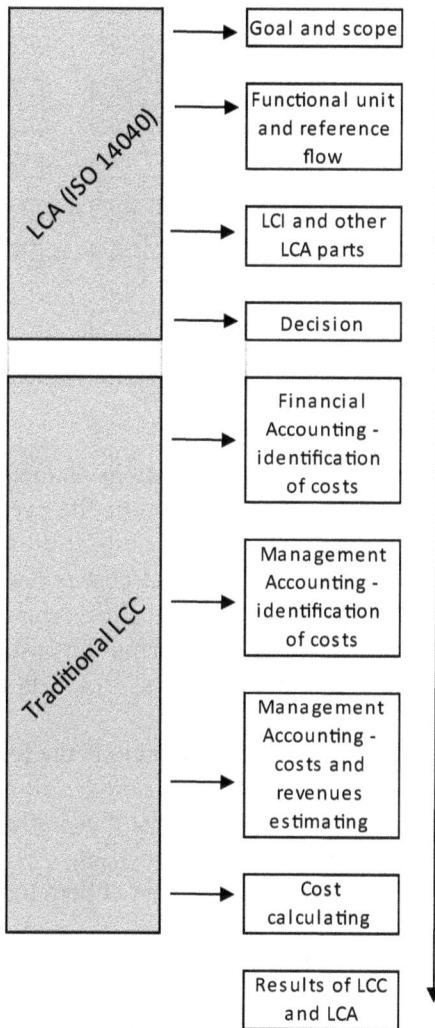

Figure 1. Procedure for LCA–LCC as part of CSR based on [16].

While LCA investigations focus on potential environmental impacts, LCC on the other hand concerns long-term economic benefits. There are some differences between these two techniques, i.e. input–output data flow on the complete set of upstream processes is necessary for the calculation of the total environmental impacts in LCA. However, LCC is concerned about cost (e.g. market price of products and services, instead of

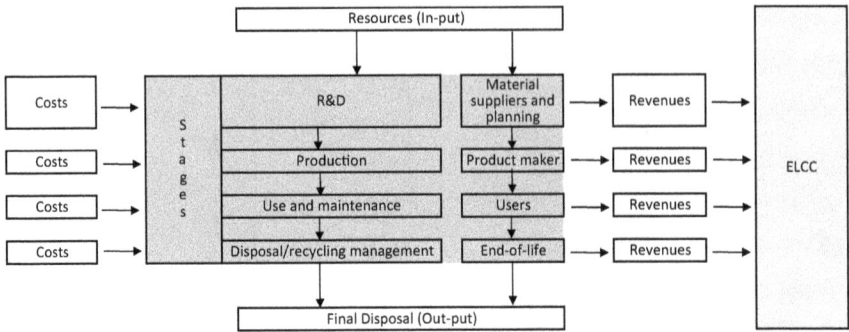

Figure 2.   Conceptual framework of fundamental Environmental LCC.

environmental or pollution concerns). A bottom-up method in LCC (total cost summed up by cost of each step during the life cycle of a product) is compiled in Table 1.

Converting LCA impacts into financial cost is one of the means of integrating both methodologies together. In this sense, LCC and LCA variables are taken into account in quantifying the cost of environmental impacts. In the most straightforward terms, ELCC is a combination of LCC and LCA.

The main objectives of this approach include the following:

- comparisons of product life cycles against those offered by alternative designs or by similar products already on offer,
- identification of direct and indirect drivers of both traditional and environmental costs,
- determination of external and internal environmental costs arising in the context of a product on offer,
- tracking and analyzing the effects of improvements introduced with reference to a product on offer,
- qualitative and quantitative estimation of changes planned with reference to a product, including changes in product life cycle and product innovations,
- identification of win–win scenarios and striking a balance between the environmental, economic, and social aspects over the course of a product life cycle; determination, calculation, and analytical evaluation of the environmental impact of both the production process and the product presented in pecuniary value [17].

Table 1.   Comparison of LCA and LCC.

	LCA	LCC
Time horizon	Average life of products	Average life of products
Method	Quantity	Quantity
	Top-down	Bottom-up
Focus	Future environmental impact Environmental impacts	Future economic value Economic value
Scope	Materials and products	Cost
Evaluated environmental impacts	Air, soil, and water	Not considered
Cost calculation	Not considered	Totally system cost

Decisions involving introduction of the basic form of ELCC should be made considering the fact that limitations and boundaries of the production system are not necessarily equivalent to those employed in LCA concept. For this reason, the standard LCA techniques should be supplemented by economic and social aspects. ELCC must be implemented in coordination with a valid concept of environmental evaluation (such as LCA) and on the grounds of an effective analysis of the social impact of products under examination within the same context of identified systemic limitations and confines. In this largely simplified and utopian scenario, all the remaining external effects are processed by other instruments typically employed for standard sustainable development evaluations. It may also be suggested that — assuming a fair construction of both fiscal and environmental subsidies systems (i.e. unbiased and fully justified by the complete set of environmental effects generated in association with a given product and reflected in the product's fixed environmental cost) — the observations related to the narrow context of an economic system may serve as a generalized fundament for detailed evaluations of both the social and the environmental aspects.

The life cycle of a product in the ELCC approach should be perceived equivalent to the physical life cycle of a functional unit. Following the approach used in the LCA technique, it may be postulated that the term should include the phases of production, operation, consumption (wear), and closure and be accompanied by the so-called knowledge phase (R&D or knowledge purchased through supply chains), complete with product utilization and its functional utility for consumers. Within the framework of the environmental life cycle costing approach, it is assumed that — for the majority of mass industrial products — the volumes of resource consumption and emission associated with the R&D phase are largely negligible in terms of their environmental effect and therefore may be analyzed for the entire assortment of products. The same applies to material and energy flows attributable to the R&D phase — these are low, since the main focus at this stage is placed on conceptual work, modelling, calculations, ideations, and laboratory work (this may not necessarily apply to large-scale projects). At the same time, it must be noted that the R&D phase exerts a direct impact upon the remaining phases of a product life cycle, and the environmental effects are no exception in this context [17].

While comparing LCA and ELCC, it may also be observed that the former provides no account of marketing activities or any technical wear and tear concerns. This limitation stems from the assumption that the above elements are insignificant from the viewpoint of LCA objectives and premises due to their typically negligible imprint in environmental calculations. ELCC, in contrast, provides a full and comprehensive account of the above aspects, both in its assumptions and in relation to the physical life cycle of a product. It should thus be considered whether their introduction would imply any cost effects or environmental footprint and if such effects are of any practical value for calculation purposes. If their significance or impact is below the 5% margin, they may safely be omitted to simplify ELCC calculation tasks. It may be postulated that any additional elements of interest from the economic viewpoint (and not necessarily in direct relation to the environmental aspects) be included in calculations unless their effects are below the pre-established margin, to allow for a better unification of the two techniques. A similar postulate can be made with reference to practical evaluation of the environmental and economic effects of production decisions. Any element of ELCC removed from economic calculations (for reasons of being deemed insignificant) may be included in the evaluation of environmental impacts and vice versa. In other words, while systems of the economic and the environmental evaluation may differ — also in terms of their boundaries (sets of the applicable forms of economic or environmental impact) — the very limits of ELCC established with reference to a given product must correspond with those defined under LCA. This serves as further evidence to support the view that LCA compliant with the ISO standards of environmental management may be directly carried over to LCC, resulting in the postulated ELCC.

Naturally, the boundaries of ELCC differ from those defined for LCA, as the focus of the former is still placed on costs generated during a product life cycle and not limited to purely environmental effects. Part of such costs may be derived directly from the introduction of LCA (see Table 2; items marked in bold italics). Calculations of their pecuniary values (determinations of specific costs) may be done by multiplying the values of flows with the monetary value of their respective cost or market price. Items marked in italics (Table 2) may be derived directly or indirectly from information contained in LCI analyses (elements of the LCA technique). This implies additional duties related to information acquisition [18]. If the phases of product R&D and marketing are removed from environmental evaluations, costs cannot be obtained from conventional LCA models and have to be determined from a different, separate evaluation.

The information compiled in Table 2 entails an effective process-based cost evaluation method, relating to both direct costs (such as material and energy flows) and indirect costs (e.g. the cost of labour).

Table 2. Connection of LCA elements with LCC elements.

	Cost for Product Manufacturer	Cost for Product User
Production	***Materials*** ***Energy*** *Machinery and factories* *Labour* ***Waste management*** ***Emission controls*** ***Transports*** Marketing activities	Purchase
Use	*Maintenance/repairs (warranty)* *Infrastructure* *Liabilities*	***Transports*** *Storage* ***Materials*** ***Energy*** *Maintenance/repair infrastructure*
End of life	***Waste collection, disassembly, recycling, and disposal***	***Waste collection*** ***Disassembly*** ***Recycling*** ***Waste disposal***

Consequently, the ELCC methodology should be expanded to cover for the following cost items:

- cost incurred in relation to physical and immaterial processes, deemed significant from the viewpoint of the product's environmental impact,
- cost deemed significant in relation to the objectives and the scope of environmental impact evaluations determined by LCA in the input-output aspect,
- social costs,
- the external environmental costs.

As suggested by the above information, the fundamental property offered by ELCC is the correlation between the process system of LCA methodology, the product, the existing material and energy flows, and a number of other variables. Consequently, ELCC terminology is based on more or less the same terms and notions as LCA concept and may safely be applied to both products and services, as the calculation methodology is the same in both cases. An example of a cradle-to-gate production system, starting from fossil fuel extraction to refinery, other material production stages, final product manufacture, use and disposal with costing factors associated with each LCA stages, is illustrated in Figure 3. Simplified ELCC factors involved are external environmental costs of pollution generated.

The combination of I/O-based and process-based LCA, also known as hybrid LCA or I/O-LCA, may prove to be of enormous value in environmental life cycle costing [19]. Hybrid LCA typically employs two input–output methods: the quantitative model and the Leontief price model. The former may be used for calculations of the so-called sectoral production level and the correlated environmental load associated with a given production output. The price model, on the other hand, allows for the calculation of prices for output products against specific indicator levels, with consideration of their added value or the estimated cost per unit of production [20]. In practice, and from a mathematical standpoint, both models (quantitative vs. price) are treated as separate and independent [21].

From the ELCC standpoint, the physical relations between inputs and outputs, in a processual approach, improve the utility of the quantitative model in LCA as basis for the investigation of costs identified in the price

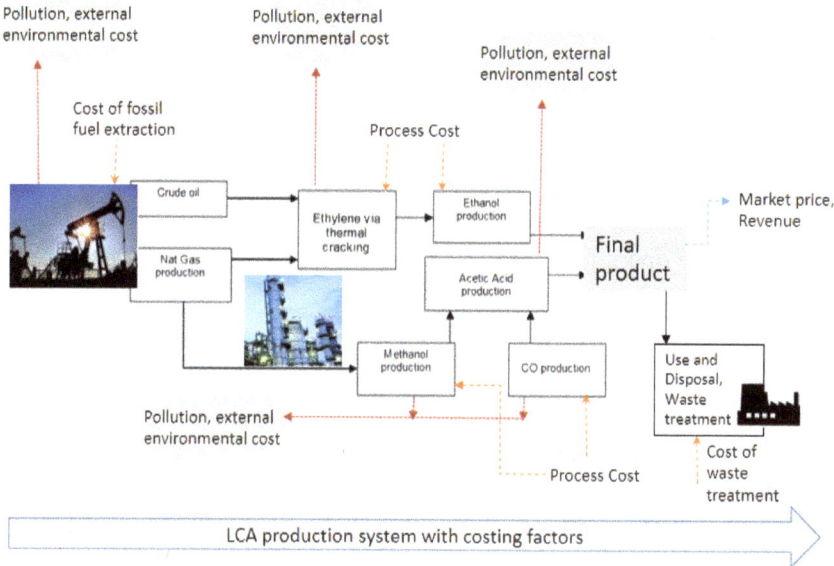

Figure 3. LCA example with ELCC costing factors.

model, while the latter places emphasis on the calculation of costs and prices. Initial determination of the input–output matrix provides a better understanding of the ELCC calculation system and is constrained by a number of principles designed to ensure cohesiveness of system operation. Of those, the most important one — at least with reference to cost analysis — is the principle stating that the corresponding values on the input and output side of the equation should be equal. Let us assume a simple scenario of a national economy comprising three industrial sectors; calculation of the above principle in the I/O matrix approach may be obtained from Equations (4)–(6):

$$p_j x_j = \sum_{i=1}^{3} p_i a_{ij} x_j + V_i, j = 1,\ldots,3, \tag{4}$$

where x_j is the production volume at input, p the price, a_{ij} the quantity of materials i required per unit j, and V the gross value added (including, among others, costs of basic products reported in sector j) [20].

The I/O model demands that the production cost per unit be equal to its market price supplemented by the gross added value per unit; this can be verified through division of both sides of the equation by a factor of x_j:

$$p_j = \sum_{i=1}^{3} p_i a_{ij} + v_j, j = 1,\ldots,3 \tag{5}$$

The above operation yields the following equation:

$$p = v(I - A)^{-1} \tag{6}$$

where p is the transposition $(p1, p2, p3)$, v the transposition $(v1, v2, v3)$, A the values obtained from the technical matrix of production elements, row i, column j, and I the unit of technical matrix No. 3 [20, pp. 10–24].

Equation (6) provides us with cost per unit or price equivalent in the I/O model approach [20]. The cost refers solely to the sum of costs generated in production phase, but may be supplemented with the annual cost of R&D and marketing. Introduction of costs generated throughout the product use phase is also straightforward, as the A value may safely be augmented by input data multiplied with time T. The resulting value represents production costs fully reflecting the character of correlations between the three sectors under examination. However, introduction of operational costs and revenues on the basis of the I/O method requires examination of issues other than those employed in the postulated equation, such as waste production and management. At this point, it may be practical to introduce a third phase representing product disposal. The output of this phase yields the volume of wastes, $p3$, which provides us with the price of waste storage and disposal calculated per unit of production. The available recycling schemes should also be addressed in calculations as prices for recycled materials and reclaimable waste [22]. A profitable resale of waste may significantly reduce per-unit costs at the final phase of product life cycle (e.g. reclamation of rare-earth metals and other elements from electronic waste; the Polish KGHM Metraco specializes in reclamation of rhenium from industrial waste).

A few LCA–LCC modeling and associated systems boundaries have been proposed [5, 7, 13]. For the purpose of identifying any economic, environmental, and social condition or eliminating any undesirable side effect of processes, the findings obtained from the ELCC method should be examined against the parallel LCA results. One of the ways to approach this is to examine the selected specific results from the LCA system against the calculated environmental life cycle costs. The combined results of these two methods can also be employed for further analyses within life cycle management (LCM) context.

Hunkeler *et al.* [17] postulated a procedure for gathering, identification, and quantification of cost data per unit process, per production subsystem or directly per product, followed by aggregation of such costs for each LCC phase. This approach involves the following steps:

- identification of subsystems or unit processes capable of generating costs or revenues,
- allocation of specific costs or prices to specific product flows in unit processes or subsystems identified in step 1 above, complete with determination of intermediate products,
- identification of cost or price differences for unit processes and subsystems identified in step 1 offered by alternative approaches to the project at hand,
- allocation of costs or prices identified in step 3 with process output as a unit of reference,
- calculation of costs per unit process or per subsystem, involving multiplication of costs per reference unit obtained in steps 2 and 4 in absolute process output values for the purpose of retaining the referential flow of the entire production system,
- aggregation of costs or prices for each unit process or subsystem within the entire life cycle [17].

Input data for calculations in steps 2 and 4 may be obtained from ERP class systems, coupled with LCI modelling software (e.g. SimaPro, GaBi, Umberto). The postulated stages may be viewed as contributing in the implementation of the "WELCC" method, supplemented with, e.g. environmental indicators or environmental marketing objectives.

Analytical evaluation of the results suggests a prevalent use of monetary (pecuniary) units and natural units of measurement (such as kg in carbon dioxide emissions) in the WELCC approach to evaluation of product impacts. It must be noted that such interpretations should be made not only with reference to results obtained per unit but also to any partial and comprehensive results. This reservation also applies to the social perspective, as a pillar of the sustainable development policy. A good example of the susceptibility of partial results (in this case, the environmental impact of economic data) is the specificity of environmental taxes and fees, strongly correlated with environmental policies defined at the state level. The postulated WELCC is based on the following assumptions:

- New conditions of enterprise operation imply the use of new approaches to cost calculation.
- New types of clients/recipients/external stakeholders expect companies to support coordinated activities within the framework of state environmental policies and principles of corporate social responsibility.
- Calculation of costs per product should properly reflect any external costs (environmental or social); enterprises should ensure proper separation of their environmental costs at least in the two basic dimensions.
- The economic, social, and environmental areas of enterprise involvement should not be examined in separation from one another — they constitute an integral whole.
- The practical attempts involving the formulation of separate reports and overviews of environmental costs as part of the obligatory financial disclosure duties should be praised as effective.
- In the foreseeable future, it seems necessary to introduce additional obligatory reports including financial and non-financial disclosures corresponding to the social, economic, and environmental dimensions of production.
- It is advisable to ensure good quality of non-financial information as well as its comparability and utility for intra-sectoral analyses.
- It is advisable to seek methods and formats for the unified presentation of environmental and social costs within the adopted boundaries of a given product system.
- Non-financial information is not easily measured, but it is advisable to make attempts to gauge their value, based on cooperation between various structural divisions of the entity.
- In practical applications, the environmental calculation of life cycle costs of a product is largely dependent on the specificity of the product, the processes, or the entity as such; however, there are more generalized, model approaches to this task readily available for implementation, regardless of the operational specificity of entities.
- The postulated WELCC model offers potential for effective use of external information by entities for operational management purposes [24].

The postulated costing model provides a set of interrelated data and can be described as a logical whole designed to reflect the environmental and cost-related impacts exerted by the production processes and the product (in each phase of its entire life cycle) and to explore the perspectives of their management in the foreseeable future. In this sense, the postulated formula employed in the WELCC method can be expressed as follows:

$$WELCC = DPC + RTC + SC + PC + SMC + RMC$$
$$+ OC + PLC + PDC + EEC + SC \qquad (7)$$

where WELCC denotes the total cost associated with a product life cycle, DPC the design and planning costs, RTC the research and testing costs, SC the supply costs, PC the production (service rendering) costs, SMC the sales and marketing costs, RMC the repair and maintenance costs (as borne by the producer/provider), OC the other costs generated in the market phase, PLC the product liquidation costs, PDC the product discontinuation costs, EEC the external environmental costs, and SC the social costs [24].

Table 3 presents an overview of the structural organization of the postulated WELCC method.

The implementation of WELCC is initiated (stages I, II, and III) by managerial evaluation of the present company market position in the environmental aspects (also in relation to products and technologies). These activities should be followed by the specification of company environmental objectives and selection of effective strategies to that effect. The managerial team should ensure that the entity is ready to face the task of integrating the economic, political, and social spheres of its operation with the demand of preserving the environmental balance and the requirement of satisfying the fundamental needs of its stakeholders, employees, and the society at large. If such readiness can be ascertained, the company may proceed with the task of designing and incorporating an effective CSR strategy in a microeconomic dimension. Analyses of life cycle phases may serve as the basis for the design of LCA. Separation of phases allows for the evaluation of environmental aspects in relation to specific design and production processes, particularly with regard to material and resource flows and waste production/emission, but also with regard to the environmental impact of the product itself. It also offers potential for assessment of any pro-environmental improvements. Even at this early stage, proper identification of phases offers good

Table 3.   Overview of the postulated WELCC method.

Stage I	Determination of the company market position in relation to the environmental aspects of its operation
Stage II	Policy of sustainable development and the CSR
Stage III	Application of controlling and financial/accounting systems in environmental management
Stage IV	Determination of product life cycle phases (both those already on offer and those in planning)
Stage V	LCA techniques for existing and anticipated products
Stage VI	Identification of costs at each life cycle phase
Stage VII	Calculation of costs per phase
Stage VIII	Environmental indicators per phase
Stage IX	Environmental report per phase
Stage X	Summary of all costs
Stage XI	Global report

potential for modernization and optimization of manufacturing processes (both for products already on offer and those in planning) and formulation of logistic strategies. They may also be used to support eco-label applications. Shifts in the product life cycle phase are directly reflected in the evaluation of technical and technological, economic, social, and environmental conditions in relation to the present condition of the market. The stage of phase determination is followed by the LCA stage — this set of activities should be performed in the most straightforward fashion, with results provided in a form comprehensible to all recipients. Proper implementation of LCA as a constituent element of the postulated WELCC model provides good potential for separation of resources, processes (both internal and external), and input–output flows attributable to specific products.

5. Conclusion

ELCC methodology proves useful for the evaluation of progress in pro-environmental activities due to the quantitative approach to the estimation of the environmental impact of production. Implementation of sustainable development principles at the company level without proper determination of indicators and threshold values (barriers or environmental limits) will not be effective. In effect, any improvement attempts may be seen as futile. In practical applications, it is assumed that the purely environmental aspects are recognized by the LCA method. However, this approach fails to incorporate the economic and social aspects, which are expected to be determined by other means. The ELCC method is formulated as a comprehensive combination of all these aspects. The main postulate posed in this context is that the implementation of ELCC methodology should be based on data collected for products already on offer, tested and verified in terms of their economic and financial profitability.

The ELCC approach is not meant to replace the established methods of cost accounting and cost management in a traditional sense of the term, despite the comparative and systemic value of the method and its emphasis on decisions made in the strict context of sustainable development. Instead, the method should be perceived as a comprehensive, well-defined, and unified instrument. As such, the method can be used to identify significant differences between product alternatives, thus offering support for cash flow decisions. ELCC may also serve to determine company potential for product improvements and modernizations during its effective life cycle, with strong focus on the environmental effects of the product itself and of the manufacturing processes employed.

Methodological problems may arise in the context of ELCC implementation in relation to other disciplines, such as the macroeconomic cost/benefit analysis; some of them may contribute to the increase of the ELCC implementation cost as such. However, the use of the LCA technique as the basis for the implementation of ELCC seems advisable and will bring measurable effects in the environmental contexts of company operation.

The postulated WELCC method may be used as a blueprint for the implementation of qualitative functions, based on the assumption that challenges related to the proper integration of processes in their environmental, social, and economic aspects are crucial for the realization of effective protection of both the environment and social health and security. At the same time, patterns of product development can be formulated on the basis of phase product development.

The postulated WELCC model places well ahead of the present demands for effective decision-making support techniques, instruments, and systems in both the short-term and long-term perspective and offers a complete representation of all aspects required in the studied context (economic/financial, social, and environmental) in accordance with the demands of sustainable development policies and other initiatives (also international and global) designed to constrain and reduce the negative environmental effects of production. Proper implementation of the WELCC model allows for the effective estimation of costs related to product manufacturing processes that may safely be borne without the risk of passing the formally defined environmental limits as expressed in the cost of emission control and the cost related to depletion of non-renewable resources (both fuels and other natural resources).

References

[1] S. Czaja and A. Becla, *Ekologiczne Podstawy Procesów Gospodarowania* (Wydawnictwo Akademii Ekonomicznej we Wrocławiu, Wrocław, 2002), p. 249.

[2] BMU-BDI. *Nachhaltigkeitsmanagement in Unternehmen, Konzepte und Instrumente zur nachhaltigen Untemehmensentwicklung* (Bundesministerium fur Umwelt, Naturschutz und Reaktorsicherheit, Berlin, 2007), p. 21.

[3] BMU-Econsense — CSM, *Nachhaltigkeitsmanagement in Unternehmen* (Berlin, 2007). http://www.bdi-online.de/Dokumente/Umweltpolitik/nachhaltigkeitsstudie.pdf.

[4] D. Coiante and L. Barra, Practical method for evaluating the real cost of electrical energy. *Int. J. Energ. Res.* **19**, 159–168 (1995).

[5] D. Koplow and J. Dernbach, Federal fossil fuel subsidies and greenhouse gas emissions: A case study of increasing transparency for fiscal policy. *Annual Rev. Energ. Environ.* **26**, 361–389 (2001).

[6] R. Hoogmartens, S. Van Passel, K. van Acker, and M. Dubois, Bridging the gap between LCA, LCC and CBA as sustainability assessment tools. *Environ. Impact Assess. Rev.* **48**, 27–33 (2014).

[7] V. Moreau and B. P. Weidema, The computational structure of environmental life cycle costing. *Int. J. Life Cycle Assess.* **20**, 1359–1363 (2015).

[8] D. Haworth, *The Principles of Life-Cycle Costing* (Industrial Forum, 1975), pp. 13–20.

[9] S. Ansari and J. Bell, *Target Costing: The Next Frontier in Strategic Cost Management* (lrvin, 1997), pp. 15, 44–45.

[10] D. G. Woodward, Life cycle costing — Theory, information acquisition and application. *IJPM* 15(6), 336 (1997).

[11] G. K. Świderska, *Informacja zarządcza w procesie formułowania i realizacji strategii firmy — wyzwanie dla polskich przedsiębiorstw* (Difin, Warszawa, 2003), p. 202.

[12] H. A. Baussabaine and R. J. Kirkham, *Whole Life-cycle Costing. Risk and Risk Responses* (Blackwell Publishing Ltd., Oxford, UK, 2004), pp. 7–9.

[13] G. Pernilla and B. Henrikke, The life cycle costing (LCC) approach: A conceptual discussion of its usefulness for environmental decision-making. *Build. Environ.* **39**, 571–580 (2004).

[14] M. D. Bovea and R. Vidal, Increasing product value by integrating environmental impact, costs and customer valuation. *Res. Conserv. Rec.* **41**, 133–145 (2004).

[15] P. R. S. da Silva and F. G. Amaral, An integrated methodology for environmental impacts and costs evaluation in industrial processes. *J. Clean. Prod.* **17**, 1339–1350 (2009).

[16] E. M. Schau, M. Traverso, A. Lehmann, and M. Finkbeiner, Life cycle costing in sustainability assessment — a case study of remanufactured alternators. *Sustainability* **3**, 2268–2288 (2011). doi: 10.3390/su3112268.

[17] D. Hunkeler, K. Lichtenvort, and G. Rebitzer, *Environmental Life Cycle Costing*, 1st Ed., (Taylor & Francis, Routledge, 2008).

[18] G. Rebitzer, Integrating life cycle costing and life cycle assessment for managing costs and environmental impacts in supply chains. In: S. Seuring and M. Goldbach (Eds.) *Cost Management in Supply Chains* (Physica, Heidelberg, 2002). doi: 10.1007/978-3-662-11377-6_8.

[19] S. Suh and G. Huppes, Methods in life cycle inventory (LCI) of a product. *J. Clean. Prod.* **13**, 687–697 (2005).

[20] R. E. Miller and P. D. Blair, *Input-Output Analysis: Foundations and Extensions* (Cambridge University Press, Cambridge, 2009), pp. 10–24.

[21] R. Dorfman, P. Samuelson, and R. Solow, *Linear Programming and Economic Analysis* (McGraw-Hill, New York, NY, 1958).

[22] S. Nakamura and Y. Kondo, A waste input-output life-cycle cost analysis of the recycling of end-of-life electrical home appliances. *Ecol. Econ.* **57**, 494–506 (2005).

[23] M. Biernacki, *Środowiskowy rachunek kosztów cyklu życia* (Wydawnictw Uniwersytetu Ekonomicznego we Wrocławiu, Wrocław, 2018), pp. 231–234.

Chapter 10

Sustainable Competitive Advantage: A Leap Forward in Sustainable Strategy with Blockchain-Enabled LCA

David TEH

Monash Business School and Faculty of Information Technology, Monash University, Melbourne, Australia

david.teh@monash.edu.au

Progress in achieving UN SDGs poses various challenging elements. While CEOs acknowledge the opportunity for competitive advantage through sustainability, only 48% are implementing sustainability into their business, and merely 21% feel business is currently playing a critical role in contributing to the SDGs. This chapter explores the potential of blockchain-enabled LCA and suggests a potential framework of how blockchain-enabled LCA may be implemented. The framework offers a step closer to unlocking the benefits of inventory data management required in LCA applications for various industries. Blockchain-enabled LCA can provide better data integrity, traceability, transparency, and management with multi-stakeholders along a global supply chain network.

1. Introduction

The application of life cycle assessment (LCA) has become widely recognized as an environmental management tool and is increasingly being applied to help with strategizing sustainability strategy, sustainability goals, and decision-making in various industries. However, LCA applications may face multiple challenges in obtaining essential data requirements and the management of information flows. As many businesses are driven to create more substantial impacts for sustainable development and achieve the United Nations (UN) Sustainable Development Goals (SDGs), the adaptation of conventional LCA frameworks to include new technological innovations is called for. While CEOs acknowledge the opportunity for developing competitive advantage through sustainability, only 48% are implementing sustainability into their business operations, and just 21% of CEOs feel business is currently playing a critical role in contributing to the SDGs [1].

Sustainability and climate change can affect a business's financial performance [2], competitive strategy [3], and long-term value creation and growth [4]. This chapter aims to revisit how blockchain technology can address the challenges companies face in inventory data management and requirements in the implementation of LCA [5, 6]. Specifically, this research seeks to further extend and demonstrate how blockchain can be integrated into LCA to address the challenges associated with *integrity*, *assurance*, *traceability*, and *transparency* of data and limitations inherent in conventional LCA, given that many stakeholders are involved in complex and interdependent activities along the value chain.

2. Background and Related Work

2.1. *Sustainability and sustainable development*

Corporate sustainability is built on the notion of sustainable development and the business approach that fosters company longevity through the development and implementation of aligned business and sustainability strategies that aim to create long-term sustainable value. This is achievable by taking into consideration how the organization operates within the ecological, social, and economic parameters.

The 2030 Agenda for Sustainable Development and its 17 SDGs (Figure 1) aim to achieve a better and more sustainable future for all by

Figure 1.   17 UN SDGs.

tackling global challenges in all countries [7]. This adds another layer of challenge to the existing corporate sustainability. First, for transformation at a large scale, multi-partnership and active collaboration of individuals, governments, public and private sector organizations are required. Since there are many different stakeholders involved, organizations' profit-maximizing strategy needs to be revised and rethought. An organization's economic success ultimately depends on the creation of monetary value and sustainable values to satisfy the broader stakeholders' needs and expectations [8] in order to remain viable and competitive [9, 10].

Previous research [11, 12] has shown a clear vision with long-term strategic focus, supported by deliberate planning and direction setting which is required for effective sustainability strategy development and implementation. This also distinguishes successful organizations from other organizations [13].

2.2. *Strategizing sustainability with blockchain technology*

This chapter refers to the framework (Figure 2) developed and discussed in [14]. Blockchain is included as part of the interconnected information systems that support business functions and internal processes. Based on an internal point of view, it would always be beneficial to explore deeper

Product-Focused Activities

	ACQUIRE	CREATE	DELIVER	SUPPORT	RECOVER	

Strategic Dimensions & Focuses

STRATEGY
STRUCTURE
SYSTEMS
including Blockchain Technology
SHARED VALUES
RESOURCES (incl. SKILLS)
STAKEHOLDERS (incl. STAFF)
STYLE

PERFORMANCE
IN
BUSINESS
&
SUSTAINABILITY
&
SDGs

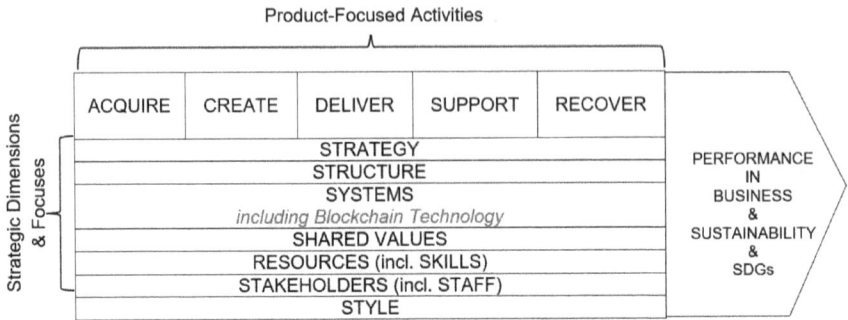

Figure 2.   Strategizing sustainability with blockchain technology for SDGs.

into the business operations and the manufacturing processes of an organization that spread across all interdependent activities along the supply chain — this can help identify "hotspots" in an LCA supply chain and hence pursue improvements for better environmental performance [15].

It is these processes which utilize (hopefully not waste) the resources that can create detrimental environmental impacts such as immense amount of waste generation, water and air pollution, and negative impacts on biodiversity. Companies who are in the materials and industrials sector, for example, mining or chemicals production companies can consider from an innovation perspective for business improvement and sustainable development. They can play a role in achieving the SDGs — SDG 9, 12, 13, and 15 [14].

On the supply side of the equation, the key medium of the impact of an organization takes place via multiple external connections with other organizations, that being the goods/services which are provided by these organizations (involve different tiers of suppliers) before reaching its end customers, which can be highly environmentally sensitive. Part of the framework (Figure 2) recognizes that the dimensions are the ubiquitous spread of the interconnected *product-focused activities*, namely to *acquire, create, deliver, support*, and *recover*, supported by the 7S within the framework.

Many organizations have started incorporating different aspects of sustainability into their business, processes, and products [15]. The benefits are many, exploring the potentials of product (re)-design [16] or identifying other material efficiency options [17].

To further facilitate and support organizations building a sustainable competitive advantage and moving toward achieving the SDGs, there is a need for more systematic and robust tools to manage sustainability performance and identify solutions that best support sustainable development and help in the decision-making process [14, 18].

2.3. *Life cycle assessment*

LCA is a methodology that enables a company to study and estimate the potential environmental, social, and economic impacts of a product/ service over its entire life cycle [19, 20]. LCA considers the entire life cycle — from the extraction and acquisition of raw materials or resources (cradle), through production to use and finally the product's end of life treatment or disposal (grave) [21–23].

Although LCA is a powerful environmental management tool, there are some inherent challenges faced in data acquisition in its application, see [24, 25]. Approximately 70–80% of the time and cost of an LCA can be attributed to data gathering for the inventory analysis phase [26], especially with a comprehensive LCA study, due to the complexity involved in the information gathering process that is appropriate for LCA modeling [27, 28]. Further, it can also be extremely expensive when data are not readily available or if one's knowledge about specific processes, materials, and emissions under study is partial [24].

In the final stage of LCA, a set of environmental impacts is generated, including global warming potential, hence providing important quantified analysis for policymakers as well as strategic planning for sustainable green companies. The reliability of such results highly depends on the operational parameters or process modeling applied in manufacturing systems, supply chain choices, and associated data/information gathered from the series of production chains within the LCA model [15, 27, 28]. While evaluating the overall LCA performance, any potentially accumulated misinformation within any stage of the supply chain can be problematic to fix or prevent; these errors that occur in the first place are regarded as the "unknown" [29, 30]. Henceforth, blockchain applications have the potential to eliminate any misguided information or data inaccuracy in future LCA applications.

2.4. *Blockchain technology*

Blockchain can offer some solutions where it can validate information digitally in near real time, facilitate coordination, and streamline activities of various stakeholders involved along the supply chain, effectively and efficiently. It enables a new way of organizing, storing, and transferring data, by decentralizing it. With this feature, it increases data security and improves data access [31–33].

More than 65 ways of blockchain can be applied to the world's most pressing environmental challenges [34]. Due to its unique characteristics and features, blockchain can bring all stakeholders together, from the acquirer, creator, intermediaries, to end consumers in LCA. The technology can potentially improve the transparency, assurance, and sustainability of *product-focused activities* within LCA methodologies to a certain extent, by sharing *immutable* and *irreversible* information in real time to meet reliable information requirements and decision-making processes [14, 29].

Blockchain is defined as "a distributed ledger technology in the form of a distributed transactional database, secured by cryptography, and governed by a consensus mechanism" [35]. Blockchain, also known as distributed ledger technology (DLT), consists of a geographically distributed peer-to-peer network of computer nodes each of which maintains a decentralized shared database or "digital ledger" of records or transactions. These distributed ledgers are formed and maintained with a verification process (called "mining") consistent with the network consensus rules [33]. It uses cryptography to keep exchanges secure, where the computer nodes must all approve an exchange or event before it can be verified, recorded, stored, and distributed on the ledger for anyone within the network to access — hence, it provides transparency. These transactions are verified and updated in near real time by consensus of participants in the system network, without the need for intermediary or central verification [36, 37]. The algorithm or distributed consensus protocol is one of the important distinct features of blockchain that provides a secure mechanism for electronic collaboration without relying upon a central intermediary for trust [38].

The peer nodes agree on the validity and the sequence of transactions. As each transaction occurs, it is placed into a data structure called "block". Each block is being connected to the previous block (the first block being the Genesis block) and added to the next in an irreversible chain and transactions within the respective blocks are connected to the shared database to form a connected chain, hence the name "blockchain". Each block in the blockchain has its own timestamp and a cryptographic hash that connects the new block to the previous block (Figure 3). Hence, blocks can

Genesis Block	Block 2	Block 3	Block N
Hash 0 Timestamp	Hash 1 Timestamp	Hash 2 Timestamp	Hash 3 Timestamp
(Data – 1)	(Data – 2)	(Data – 3)	(Data – N)
Tx1 Transaction 1 ... Txn Transaction-n	Tx1 Transaction 1 ... Txn Transaction-n	Tx1 Transaction 1 ... Txn Transaction-n	Tx1 Transaction 1 ... Txn Transaction-n
Hash 1	Hash 2	Hash 3	Hash N

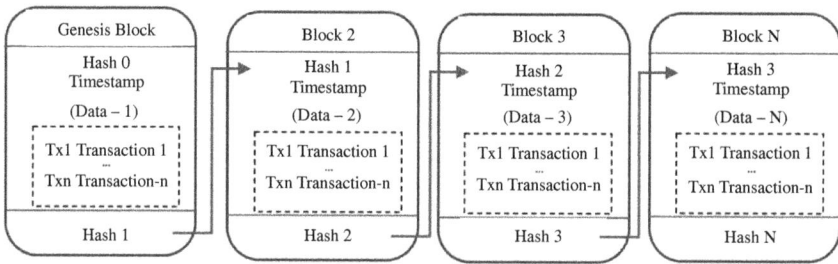

Figure 3.   How blockchain works.

only be appended, not deleted which make the transactions difficult, if not, impossible to be manipulated [36, 37].

In a nutshell, the shared database will continue growing with more records being added to the existing blockchain. Because those records are *immutable* and *irreversible*, tampering of any block information can be detected by any peer nodes on the blockchain. Each of the record stored in the block is time-stamped and encrypted utilizing a rigorous verification process, therefore, no individual entity can change or alter the verified and recorded information in the block that is being connected to the previous block, with information available throughout the distributed ledger network [33].

Since it is a write-once, append-many technology, it enables each transaction to be verifiable and auditable. Blockchain enforces transparency, provides assurance, and guarantees the ultimate system-wide consensus on the validity of an entire chain of transactions history [39]. Technical details of blockchain are discussed in detail by [40–42].

3. Blockchain-Enabled LCA for Sustainability Strategy and SDGs

The main characteristic of blockchain which can serve an extremely beneficial purpose for sustainability in LCA is its transparency and open nature. Using blockchain, LCA of products/services can be completed using actual real-time data collected via Internet of things (IoT) in LCA methods [43]. With blockchain-enabled LCA, products can be tracked across global supply chains and specifically, for the inventory analysis stage, corresponding environmental and/or social impacts can be clearly identified and verified [44]. It is a revolutionary contribution of blockchain to the tracking of reliable and real-time information in the LCA domain [29, 45].

Further, blockchain has the potential to help enforce accountability and govern ethical behaviors on behalf of the various businesses involved in the supply chain [46]. For instance, some companies in the materials industry groups have been criticized on numerous occasions due to the use of raw materials (natural resources and minerals) that are unsustainable and the associated social and environmental impacts of products developed and sold by these companies as well as accountability-related issues pertinent to the production and business operations along the supply chain [14]. This could potentially lead the companies to making unintended false claims, where the information cannot be easily verified, which can increase companies' reputation risk.

Blockchain can address the challenges associated with *integrity*, *traceability*, and *transparency* of data for future LCA applications. This may simplify the data gathering process across the LCA, where multidimensional and multi-domain data at multiple locations (IoT-enabled) can be collected, specifically for the inventory analysis step in the LCA.

Blockchain should be considered and used by companies who operate in the material sector or environmentally sensitive industries that undertake a complete LCA to help understand their own product/service life cycle and the associated impacts on the environment and society and facilitating the achievement of SDG 9, 12 and 13. To a greater extent, blockchain can help companies to track the origins of raw materials, and movements of products; inventories can be tagged with related environmental and social information such as results of environmental product declarations and greenhouse gas (GHG) emissions accounting information [47, 48] to possibly avoid the potential environmental burden or minimize trade-off [21, 22]. This, for example, can help in achieving multiple targets under SDG 12, including SDG 12.2 sustainable management and efficient use of natural resources, SDG 12.4 environmentally sound management of chemicals and all wastes throughout their life cycle, SDG 12.5 substantially reduce waste generation through prevention, reduction, recycling and reuse, and SDG 12.8 promote universal understanding of sustainable lifestyles.

Blockchain can be used to advance the attainment of other SDGs, for example, climate change (SDG 13), biodiversity and conversation (SDG 15), healthy oceans (SDG 14), water security (SDG 6), clean air and weather (SDG 13), and disaster resilience (critical for SDG 1; can be contributed by SDG 4) [34]. These global environmental challenges have significant detrimental impacts on both the economy and society. Blockchain can be a powerful technology to support all the stakeholders involved in the value chain. In the context of corporate sustainability, blockchain-enabled solutions are expected to improve the reliability of data related to supply chains and to help businesses eliminate waste and hazardous activities [34].

There are multiple applications of blockchain in product life cycle or supply chain, for example, authenticating the reporting of carbon emissions, energy use and intensity, pollution, water, and waste, hence verifying suppliers' compliance with environmental sustainability requirements and standards. This can also reinforce the integrity and value of certification for better informed customers' decisions and selection of products [14, 49].

Besides, blockchain can be used to improve organization's internal controls system such as sustainability-related controls. For instance, implementing management systems and internal key performance indicators relating to sustainability such as the intensity of GHG emissions, the number of environmental and/or social issues along the supply chain as well as the managing company's sustainability certifications and sustainability-related risks and issues [50].

How blockchain-enabled solutions can potentially be applied together with LCA? At the beginning phase, *Acquire*, the Genesis block is used to start, some of the data such as proof of fair payment for raw materials acquisition, Scope 2 or 3 GHG emissions will be recorded in the Genesis block. This block will be time-stamped and encrypted. All relevant stakeholders in the network will verify this block before it is appended to the next block in the chain, i.e. Block 2. Now, Block 2 is created and added in the *Create* phase. Data such as Scope 1 GHG emissions, water consumption, energy use, and waste generated will be recorded in this block. Similarly, Block 2 will be time-stamped and encrypted, then this block is linked to the Genesis block created earlier (Block 1) and it will be appended to the next block, i.e. Block 3 and so on (Figure 4). This process will continue; each block will be linked to the previous block and appended to the next block (*Block-n*). It is because each block can only be appended, not deleted which makes the chain of transactions difficult if not impossible to be manipulated [36, 37].

A proposed blockchain-enabled LCA framework is illustrated in Figure 4. The model is adapted from the following: a blockchain-enabled

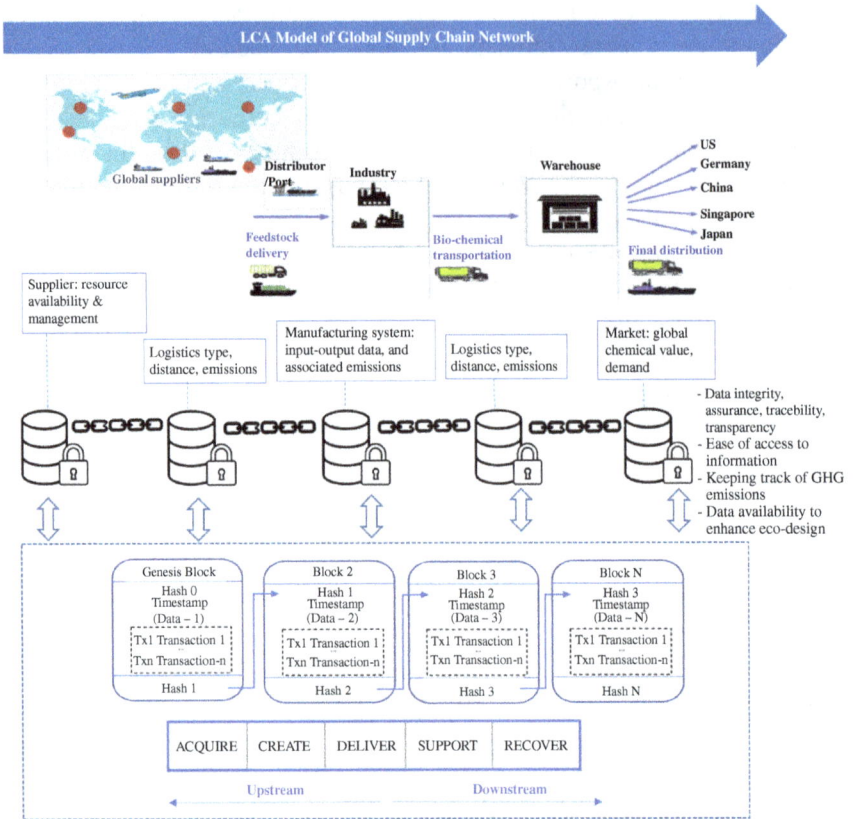

Figure 4.   The framework of blockchain application.

LCA model of production chains in a material industry [14] and a comprehensive LCA supply chain network starting with global supplies of bio-resources sent to various bio-refineries for thermo-chemical conversion to bio-chemicals, logistical analysis, and ends with the final distribution of products based on market demands [15].

It is envisaged that blockchain-enabled LCA can aid organizations aiming to improve environmental impacts of a product/service via a better understanding of the information flows within an entire production chain connecting global suppliers to manufacturing systems and final product distribution [29, 30]. Blockchain-enabled LCA provides the access to understand how the entire production/service process works. The decentralized database of relevant stakeholders (see Figure 2) includes raw material acquisition to production, logistical networks of delivery, use, and disposal.

With the support of blockchain's distinct features such as traceability and transparency, companies can build more robust and streamlined supply chain management practices to better track the products along the supply chain and its associated environmental performance [29].

Blockchain innovation has the potential to transform the inter-organizational processes in the environment and their associated activities when they are better understood — they can be mapped, streamlined, and operated [14, 51]. To this point, we have seen that there is a clear need to understand the processes of an organization in relation to each of the activities from sourcing and using resources, to the production of services and products, and then the generation of waste at the end of the product life cycle.

As suggested, the more forward-thinking and innovative organizations would use this as an unfair competitive advantage to start to rethink about product design, material use and efficiency, production capabilities, as well as the overall business model [14]. If blockchain can deliver its full potential, it can contribute to the organizations at many levels — from the strategic level for sustainability strategy development, to product innovation and creation, making it easier and faster. In addition, organizations' business practices and activities can be held accountable by their stakeholders. When these stakeholders can rely on data with higher integrity and more transparency, they become more confident with the organizations that they are buying from, dealing with, or supporting.

4. Conclusion and Future Research

This chapter extended the prior research [14] by providing more details of how blockchain works and how it can be integrated into LCA to address the *integrity, traceability, assurance,* and *transparency* challenges associated with the data and limitations inherent in conventional LCA. It has also developed a blockchain-enabled LCA implementation framework. This will support the rethinking of business operation, readjusting the business model, redesigning the products/service, and refocusing on innovative R&D activities that empower companies to be more sustainable.

The business sector is at the center of strategies for achieving the UN SDGs for 2030 because it drives the growth of the economy, while its operation can have an impact on the environment and society. Therefore, a critical step for any organization is to adopt a more long-term and strategic approach, identify how the SDGs directly and indirectly relate to their business, and align their corporate priorities with the relevant SDGs to better engage with their key stakeholders to make positive impacts.

Organizations can undertake LCA to learn and better understand their own product/service life cycle and the associated impacts on the environment and society, so that they can be more sustainable in their production, consumption, and overall business operations. To support this, blockchain can offer the solution, enabling business to facilitate the coordination and better streamline the activities of various stakeholders participating in the interconnected supply chain so that they can engage and support all the stakeholders more effectively and efficiently. This solution supports businesses who are more forward-thinking and innovative to embrace blockchain-enabled LCA as a strategic tool to develop a single source of truth of data to rely on to support their decision-making and stakeholder engagements, build their capabilities, and engineer for more sustained competitive advantage.

Blockchain-enabled LCA is capable of bringing all stakeholders together, from the acquirer, creator, intermediaries, to end consumers in LCA analysis and implementation. It can efficiently validate data digitally, facilitate coordination, and better streamline activities which enable product-focused activities to be more transparent and sustainable. It can help address the limitations of conventional LCA to a certain extent, by sharing immutable and irreversible information in near real time to meet the information requirements and decision-making process [43, 52].

However, it is also important to give some consideration to the limitations associated with blockchain studies and applications, such as infrastructure investment, risk, confidentiality issues, and costs [14]; governance mechanisms and structure [53]; non-permissioned (inaccessible data) *vs.* permissioned blockchain networks [54]; and the role of ownership including public or private access [54]. In any unfortunate cases where any stakeholder within the series of the value chains is reluctant to share valuable and critical information, the full potential of adopting blockchain-enabled LCA will be deemed unsuccessful [52, 54].

In terms of future research, further work is required to explore this exciting research domain. This includes inviting companies to participate and use blockchain-based LCA, to test how the *product-focused activities* are supported by the developed blockchain application framework. Future research may also investigate the critical success factors or barriers of implementing such a blockchain-based LCA, particularly to address some of the limitations raised in this paper. Zhang *et al.* [29] proposed a noteworthy systems architecture for "blockchain-based LCA". However, a global supply chain network with corresponding market demands [15] and details of blockchain applied for strategizing sustainability for SDGs (shown in Figure 2), which are essential for achieving sustainable development goals [7], was not mentioned.

From a technology perspective, as 5G capabilities will further accelerate innovation and support machine learning (ML) and IoT at scale by overcoming scalability issues, future research can explore more IoT-driven solutions such as IoT-driven blockchain to develop a deeper understanding of how IoT and blockchain can work together to collect multi-domain data at multiple locations that can further support blockchain-based LCA or more general blockchain-enabled solutions for supply chain management [43, 45].

In conclusion, this chapter offers another step toward providing a sound basis for future research on blockchain-enabled solutions for sustainability which have the potential to help organizations to utilize blockchain-enabled LCA that provide data integrity, better assurance, traceability, and transparency, working with relevant stakeholders along the supply chain to achieve the ambitious SDGs, collaboratively.

References

[1] Accenture. The United Nations global compact-accenture strategy ceo study — The decade to deliver: A call to business action. (2019). https://www.accenture.com/_acnmedia/PDF-109/Accenture-UNGC-CEO-Study.pdf.

[2] E. Kartadjumena and W. Rodgers, Executive compensation, sustainability, climate, environmental concerns, and company financial performance: Evidence from Indonesian commercial banks. *Sustainability* **11**(6), 1673 (2019).

[3] A. Danso, S. Adomako, J. Amankwah-Amoah, S. Owusu-Agyei, and R. Konadu, Environmental sustainability orientation, competitive strategy and financial performance. *Bus Strategy and the Environ.* **28**(5), 885–895 (2019).

[4] P. F. Dilling and P. Harris, Reporting on long-term value creation by Canadian companies: A longitudinal assessment. *J. Clean Prod.* **191**, 350–360 (2018).

[5] D. Cespi *et al.*, Life cycle inventory improvement in the pharmaceutical sector: Assessment of the sustainability combining PMI and LCA tools. *Green Chem.* **17**, 3390–3400 (2015).

[6] R. Hischier *et al.*, Establishing life cycle inventories of chemicals based on differing data availability. *Int. J. Life Cycle Assess.* **10**, 59–67 (2005).

[7] United Nations, UN adopts new Global Goals, charting sustainable development for people and planet by 2030. (2015). https://news.un.org/en/story/2015/09/509732-un-adopts-new-global-goals-charting-sustainable-development-people-and-planet.

[8] S. N. Morioka, I. Bolis, S. Evans, and M. M. Carvalho, Transforming sustainability challenges into competitive advantage: Multiple case studies kaleidoscope converging into sustainable business models. *J. Clean Prod.* **167**, 723–738 (2017).

[9] J. Elkington, *Cannibals with Forks: The Triple Bottom Line of Twenty-First Century Business* (Capstone, Oxford, 1997).

[10] C. Laszlo and P. Cescau, *Sustainable Value: How the World's Leading Companies are Doing Well by Doing Good* (Routledge, London, 2017).

[11] M. J. Epstein, *Making Sustainability Work: Best Practices in Managing and Measuring Corporate Social, Environmental and Economic Impacts* (Routledge, London, 2008).

[12] J. Estes, *Smart Green: How to Implement Sustainable Business Practices in Any Industry-and Make Money* (Wiley, Hoboken, 2009).

[13] R. J. Baumgartner and R. Rauter, Strategic perspectives of corporate sustainability management to develop a sustainable organization. *J. Clean Prod.* **140**, 81–92 (2017).

[14] D. Teh, T. Khan, B. Corbitt, and C. E. Ong, Sustainability strategy and blockchain-enabled life cycle assessment: A focus on materials industry. *Environ. Sys. Decis.* **40**(4), 605–622 (2020).

[15] H. H. Khoo *et al.*, Sustainability assessment of biorefinery production chains: A combined LCA-supply chain approach. *J. Clean Prod.* **235**, 1116–1137 (2019).

[16] S. Ahmad, K. Y. Wong, M. L. Tseng, and W. P. Wong, Sustainable product design and development: A review of tools, applications and research prospects. *Resour. Conserv. Recycle* **132**, 49–61 (2018).

[17] J. M. Allwood, M. F. Ashby, T. G. Gutowski, and E. Worrell, Material efficiency: A white paper. *Resour. Conserv. Recycle* **55**(3), 362–381 (2011).

[18] D. Teh and B. Corbitt, Building sustainability strategy in business. *J. Business Strat.* **36**(6), 39–46 (2015).

[19] M. Z. Hauschild, R. K. Rosenbaum, S. Olsen (Eds.), *Life Cycle Assessment — Theory and Practice* (Springer, New York, 2018).

[20] M. Finkbeiner, From the 40s to the 70s — The future of LCA in the ISO 14000 family. *Int. J. Life Cycle Assess.* **18**, 1–4 (2013).

[21] H. H. Khoo, LCA of plastic waste recovery into recycled materials, energy and fuels in Singapore. *Res. Conserv. Recycl.* **145**, 67–77 (2019).

[22] I. Linkov, B. D. Trump, B. A. Wender, T. P. Seager, A. J. Kennedy, and J. M. Keisler, Integrate life-cycle assessment and risk analysis results, not methods. *Nat. Nanotechnol.* **12**(8), 740–743 (2017).

[23] O. Ortiz, F. Castells, and G. Sonnemann, Sustainability in the construction industry: A review of recent developments based on LCA. *Constr. Build Mater.* **23**(1), 28–39 (2009).

[24] H. H. Khoo, V. Isoni, and P. N. Sharratt. LCI data selection criteria for a multidisciplinary research team: LCA applied to solvents and chemicals. *Sustain. Prod. Consump.* **16**, 68–87 (2018).

[25] S. Righi, A. Dal Pozzo, A. Tugnoli *et al.,* The availability of suitable datasets for the LCA analysis of chemical substances. In: S. Maranghi and C. Brondi (Eds.), *Life Cycle Assessment in the Chemical Product Chain* (Springer, Cham, 2020).

[26] J. H. Miah, A. Griffiths, R. McNeill *et al.*, A framework for increasing the availability of life cycle inventory data based on the role of multinational companies. *Int. J. Life Cycle Assess.* **23**(9), 1744–1760 (2018).

[27] N. Pelletier, F. Ardente, M. Brandão *et al.*, Rationales for and limitations of preferred solutions for multi-functionality problems in LCA: Is increased consistency possible? *Int. J. Life Cycle Assess.* **20**, 74–86 (2015).

[28] A. Zhang, R. Y. Zhong, M. Farooque *et al.*, Blockchain-based life cycle assessment: An implementation framework and system architecture. *Res. Conserv. Rec.* **152**, 104512 (2020).

[29] B. Esmaeilian, J. Sarkis, K. Lewis, and S. Behdad, Blockchain for the future of sustainable supply chain management in Industry 4.0. *Res. Conserv. Rec.* **163**, 105064 (2020).

[30] T. Jusselme, E. Rey, and M. Andersen, An integrative approach for embodied energy: Towards an LCA-based data-driven design method. *Renew. Sustain. Energ. Rev.* **88**, 123–132 (2018).

[31] M. Crosby, P Pattanayak, S. Verma, and V. Kalyanaraman, Blockchain technology: Beyond bitcoin. *Appl. Innovation* **2**(6–10), 71 (2016).

[32] R. Beck, Beyond bitcoin: The rise of blockchain world. *Computer* **51**(2), 54–58 (2018).

[33] D. Tapscott and A. Tapscott, *Blockchain Revolution: How the Technology Behind Bitcoin is Changing Money, Business, and the World* (Penguin Random House, New York, 2016).

[34] World Economic Forum, *Building Block(chain)s for a Better Planet* (World Economic Forum, 2018), https://www3.weforum.org/docs/WEF_Building-Blockchains.pdf.

[35] R. Beck, M. Avital, M. Rossi, and J. B. Thatcher, Blockchain technology in business and information systems research. *Bus. Inf. Sys. Eng.* **59**(6), 381–384 (2017).

[36] M. Iansiti and K. R. Lakhani, The truth about blockchain. *Harv. Bus. Rev.* **95**(1), 119–127 (2017).

[37] C. Shen and F. Pena-Mora, Blockchain for cities — A systematic literature review. *IEEE Access* **6**, 76787–76819 (2018).

[38] M. Swan, *Blockchain: Blueprint for a New Economy* (O'Reilly Media, Inc, Sebastopol, CA, 2015).

[39] M. Risius and K. A. Spohrer, Blockchain research framework. *Bus. Inf. Sys. Eng.* **59**(6), 385–409 (2017).

[40] V. Buterin, A next-generation smart contract and decentralized application platform. (Ethereum Foundation, 2014). https://www.ethereum.org/foundation.

[41] G. Wood, Ethereum: A secure decentralised generalised transaction ledger. *Ethereum Project Yellow Paper* 1–32, 151 (2014).

[42] P. Giungato, R. Rana, A. Tarabella *et al.*, Current trends in sustainability of bitcoins and related blockchain technology. *Sustainability* **9**, 2214 (2017).

[43] C. Favi, M. Germani, M. Mandolini, and M. Marconi, Implementation of a software platform to support an eco-design methodology within a manufacturing firm. *Int. J. Sustain. Eng.* **11**(2), 79–96 (2018).

[44] R. Shan, Y. Zhang, Y. Liu *et al.*, A comprehensive review of big data analytics throughout product lifecycle to support sustainable smart manufacturing: A framework, challenges and future research directions. *J. Clean. Prod.* **210**, 1343–1365 (2019).

[45] M. Kouhizadeh and J. Sarkis, Blockchain practices, potentials, and perspectives in greening supply chains. *Sustainability* **10**(10), 652 (2018).

[46] R. Adams, B. Kewell, and G. Parry, *Blockchain for Good? Digital Ledger Technology and Sustainable Development Goals. Handbook of Sustainability and Social Science Research* (Springer, Cham, 2018).

[47] A. Banerjee, *Re-Engineering the Carbon Supply Chain with Blockchain Technology* (Infosys Ltd., 2018), https://www.infosys.com/oracle/white-papers/documents/carbonsupply-chain-blockchain-technology.pdf.

[48] J. Sinistore, Driving sustainability through life cycle assessment, renewable energy and blockchain. (2018), https://www.wsp.com/en-AU/insights/blockchain-driving-sustainability.

[49] Provenance, Case studies. (2020), https://www.provenance.org/case-studies.

[50] J. K. Campos and T. Rebs, Opportunities of combining sustainable supply chain management practices for performance improvement. In *Social and Environment Dimensions of Organizations and Supply Chains* (Springer, Cham, 2018).

[51] J. Mendling, I. Weber, W. V. D. Aalst *et al.*, Blockchains for business process management-challenges and opportunities. *ACM Trans. Manag. Inf. Sys. (TMIS)* **9**(1), 1–16 (2018).

[52] S. A. Molina-Murillo and T. M. Smith, Exploring the use and impact of LCA-based information in corporate communications. *Int. J. Life Cycle Assess.* **14**, 184–194 (2009).

[53] B. D. Trump, M. V. Florin, H. S. Matthews, D. Sicker, and I. Linkov, Governing the use of blockchain and distributed ledger technologies: Not one-size-fits-all. *IEEE Eng. Manag. Rev.* **46**(3), 56–62 (2018).

[54] S. Saberi, M. Kouhizadeh, J. Sarkis, and L. Shen, Blockchain technology and its relationships to sustainable supply chain management. *Int. J. Prod. Res.* **57**(7), 2117–2135 (2019).

Chapter 11

Life Cycle Assessment Used for Assisting Decision-Making Toward Sustainable Businesses

**Rodrigo SALVADOR*, Murillo Vetroni BARROS†,
Diego Alexis Ramos HUARACHI‡,
Romulo Henrique Gomes de JESUS§,
Cassiano Moro PIEKARSKI¶ and
Antonio Carlos de FRANCISCO‖**

*Universidade Tecnológica Federal do Paraná,
330 Doutor Washington Subtil Chueire
St. — Jardim Carvalho — Postal Code 84017,
Ponta Grossa, Paraná, Brazil*

*salvador.rodrigors@gmail.com

†murillo.vetroni@gmail.com

‡diegorahu@gmail.com

§romulohgj@gmail.com

¶piekarski@utfpr.edu.br

‖acfrancisco@utfpr.edu.br

Life cycle assessment (LCA) has evolved into a decision-making tool, apart from serving the purpose of environmental assessments of products and services. Several aspects of businesses are influenced by LCA results, which help direct organizations toward more sustainable business conduct. In this chapter, we list key aspects of businesses that are influenced by LCA and analyze how these influences take place. It has been identified that LCA contributes to the following 10 business aspects: research and development and innovation, product development, operations management, economy, marketing and environmental labeling, corporate social responsibility, strategic planning, reverse logistics, supply chain management, and legislation and policy. Across these influences, there are challenges and opportunities in the use of LCA results within business management, which render a range of managerial implications. On the one hand, it implies continuous investments and changes in the organizational culture and values. LCA results allow businesses to optimize processes and use of resources, to reduce emissions, to achieve a better reputation, and to gain a competitive advantage.

1. Introduction

In an ever-changing global market, companies that aim to obtain competitive advantages need to meet societal demands providing products that are labeled with environmentally friendly aspects with no unintended harmful effects caused to the natural ecosystem. With regard to such claims, companies will need to prove their environmental responsibility, which ends up also triggering social responsibility traits, by providing values that go beyond the use of a product or service [1, 2].

Companies need to be entrepreneurial and innovative by adopting tools that aid in integrating environmental aspects into business results in order to provide customers with sustainable values. In this context, life cycle assessment (LCA) is a tool that stands out [3]. LCA is a methodology that assesses environmental aspects and potential environmental impacts throughout the life cycle of goods and/or services.

LCA emerged in the mid-1970s during the energy crisis to account for the consumption of material and energy as well as the related emissions [4]. In the 1990s, the use of the LCA became more apparent, assisting managers in decision-making as companies began to consider not only the use of natural resources and gaseous emissions but also the environmental consequences of the entire life cycle of products [5]. Currently, LCA studies are guided by standards from the International Organization for Standardization (ISO), such as ISO 14040 [6] that addresses the principles and structures and ISO 14044 [7] that addresses the requirements and guidelines [8] of an LCA.

LCA has been developing ever since its emergence to become in the 2020s a tool with characteristics that provide companies with a unique possibility for the detailed management of their products and processes as well as enable tracking the long-term consequences of the consumption, use, and final destination of these products. The results of an LCA can influence several aspects of businesses and can assist these businesses toward becoming more sustainable. Therefore, more than a tool for environmental assessments, LCA is also a tool to assist scientifically quantified decision-making. Based on all of the aforementioned, this chapter aims to present the contributions of LCA to key business aspects that help direct organizations toward more sustainable business conduct.

2. Methods

This chapter was based on evidence found in the existing literature for the contributions of LCA to sustainable businesses. To this end, a systematic literature review was conducted. The methodological steps taken to structure this chapter are illustrated in Figure 1.

Step 1 — Searches in databases: In this initial step, the following keyword combination was used to conduct searches in the databases ScienceDirect, Scopus, and Web of Science: *("life cycle assessment" OR "life cycle analysis" OR "LCA") AND ("sustainable business" OR "sustainable businesses")*. A total of 81 documents, including books, book sections, conference proceedings, journal articles, reports, and serials, were retrieved.

Step 2 — Excluding duplicates: All duplicate documents were excluded, which accounted for 29 documents. Thus, 52 documents remained after Step 2.

Step 3 — Filtering by title and keywords: In this step, titles and keywords in each document were read, and documents were found to be not contributing to unveiling the contributions of LCA to key business aspects

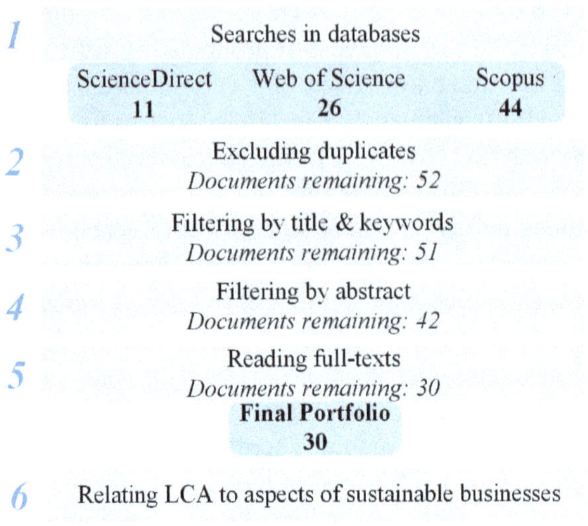

1	Searches in databases

ScienceDirect	Web of Science	Scopus
11	26	44

2	Excluding duplicates *Documents remaining: 52*

3	Filtering by title & keywords *Documents remaining: 51*

4	Filtering by abstract *Documents remaining: 42*

5	Reading full-texts *Documents remaining: 30* **Final Portfolio** **30**

6	Relating LCA to aspects of sustainable businesses

Figure 1.   Overview of research methods.

that help direct organizations toward a more sustainable business conduct were ruled out.

Step 4 — Filtering by abstract: The same procedure followed in Step 3 was followed here, but applied to the abstracts. The abstracts of the remaining documents were read and documents were found to be not contributing to unveiling the contributions of LCA to key business aspects that help direct organizations toward more sustainable business conduct were ruled out.

Step 5 — Reading full texts: The full texts of the remaining documents were retrieved and they were thoroughly examined. The pieces of research, once again, found to be not contributing to unveiling the contributions of LCA to key business aspects that help direct organizations toward more sustainable business conduct were ruled out. After these filters, the documents that remained constituted the final portfolio of this research, which comprised 30 pieces.

Step 6 — Relating LCA to aspects of sustainable businesses: The documents in the final portfolio were read, seeking to fulfill the aim of this research. Thus, the authors identified the key aspects of businesses that are influenced by LCA and what these influences are for making businesses more sustainable. These areas and influences are presented in the following section.

3. Influences of Life Cycle Assessment on Key Business Aspects Contributing to More Sustainable Businesses

The influence of LCA toward more sustainable businesses can be seen across a range of business aspects. The literature shows that the contribution of LCA to enhance more sustainable business directions is still a matter of large debate. Some aspects seem to receive greater influence than others. Research and development and innovation, product development, operations management, economy, marketing and environmental labeling, corporate social responsibility, strategic planning, reverse logistics, supply chain management, and legislation and policy are a few key aspects of businesses where the influence of LCA may contribute to greater sustainability.

Seeking to summarize the main influences of LCA over these aspects, Figure 2 presents a synthesis of the main contributions of LCA to the key business aspects.

The insights highlighted in Figure 2 are discussed hereafter.

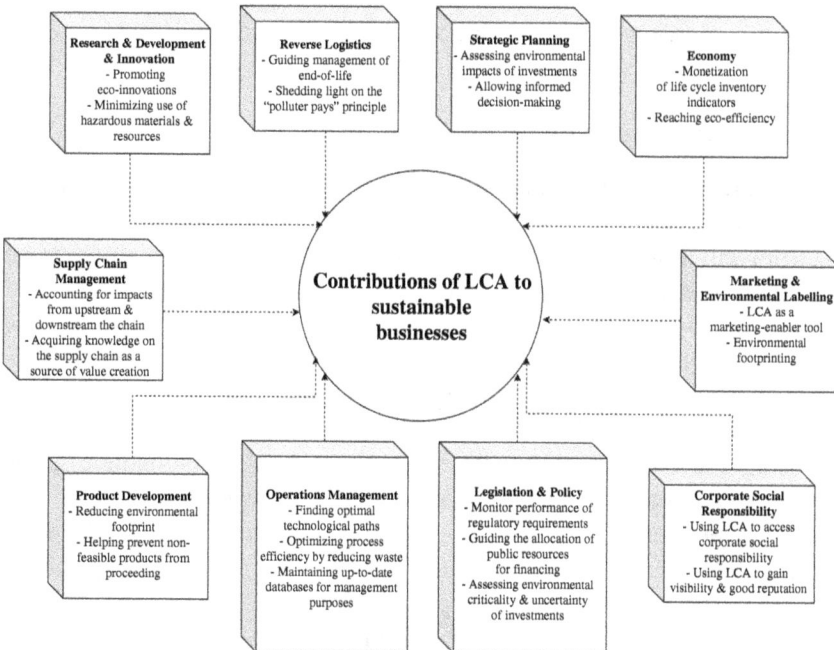

Figure 2. Contributions of LCA to sustainable businesses.

3.1. *Research and development and innovation*

Undoubtedly, LCA contributes to innovation. Indeed, an eco-innovation, as its definition points out, seeks to create products, services, processes, or systems that minimize the use of both hazardous materials and natural resources, including energy and land, throughout their life cycles [9]. In this sense, simple things such as the reuse of water or lowering energy consumption, which can be guided by LCA results, can be considered eco-innovations. The definition of eco-innovation is comprehensive, thus talking about contributions of LCA to innovation necessarily involves other processes of the company, such as product development, operations management, reverse logistics, and others.

Opportunities for eco-innovation have been found based on LCA results. Barbieri and Santos [10] report that there is a correlation between eco-innovation strategies and product and process development, and they argue that to ensure the success of these strategies, it is necessary that sustainability be understood by both management and operations areas. LCA has the potential to determine whether the implementation of eco-innovations will generate "benefits" for the environment and in which processes or stages, thus also contributing to the establishment of indicators for monitoring and improving sustainable measures [10, 11].

3.2. *Product development*

Environmentally sustainable businesses and business models are usually oriented to develop more sustainable goods or services [12]. There is a need for environmental management tools to define and monitor green efficiency metrics to provide quantified information for analyzing and optimizing associated processes within business activities properly. One of the highlighted contributions of LCA to the product development process is the possibility of anticipating the environmental impacts of a product throughout its entire life cycle (by simulating scenarios) and thus choosing the configuration with the lowest impacts. Nonetheless, the development phase is one of the most important when developing new goods or services, for most of the life cycle impacts will be analyzed at this phase [13]. In addition, LCA also helps prevent non-feasible products and processes from being set into practice [14].

Using LCA for incorporating and maintaining sustainable business practices (via, e.g. the selection of materials) has been in practice for a long time (see [15]). LCA has proven to be able to help create eco-efficient value [3], where the environmental footprint is reduced while making the business feasible, considering the customer's perceived value and the ecological/environmental costs.

3.3. *Operations management*

Moreover, from a decision-making perspective, LCA also plays a role in assisting to decide on the processes and technologies that are best or most appropriate to turn an idea into a product by revealing the technological path with the greatest potential to optimize environmental gains, either solely from an environmental perspective or coupled with economic aspects [16]. Based on LCA applications, strategic business objectives can be appropriately defined based on a set of ecological metrics (e.g. CO_2 emission, material/resource consumption, energy intensities, and wastes) to be measured [15, 17] as well as provide insights for optimizing process efficiency via reducing (or abolishing) the generation of wastes.

According to Brehmer *et al.* [12], a business or business model does not need to be sustainable to improve the efficiency of operations; in fact, projects aiming to increase efficiency in the use of material or energy are often motivated by economic reasons (e.g. cost reduction) rather than environmental ones. Thus, correct management of operations is not environmentally sustainable by itself, however, it can be a basis for a company to pursue greater sustainability. Companies might be encouraged to keep up-to-date data on their processes for management purposes and might use digital platforms/means to store and manage such datasets [2], which might also account for economic and social aspects.

3.4. *Economy*

Indicators from a life cycle inventory can be converted into economic intensity measures, thus analyzing investment and facilitating to find out whether companies are truly environmentally sustainable [18]. For investors, it is increasingly important to know whether companies are sustainable because in a greener and more resource-constrained future global context, sustainable businesses are more likely to prevail economically.

LCA results can lead to eco-efficiency and direct economic benefits, showing how life cycle costs might be reduced by replacing and/or reducing material and energy inputs or even switching technologies (e.g. through innovation, see Section 3.1) [19]. On top of that, LCA, community eco-management, and audit schemes are said to have increasingly gained the attention of society in general and hence have become a societal demand for expanding businesses [20].

3.5. *Marketing and environmental labeling*

LCA has also been referred to as a great marketing tool, since manufacturers can gauge a competitive edge in markets that demand goods and services that are more environmentally friendly [16]. In addition, LCA has been used to analyze the environmental footprints of products [17]. Unilever is an example of a company that has been relying on LCA studies to map the environmental profile of thousands of its products [21].

3.6. *Corporate social responsibility*

In ISO 26000 [22], the environment is said to be a core subject for CSR [23], and although environmental sustainability can be assessed using several techniques, LCA is a tool commonly used for this purpose [24, 25]. Its results can be used to power CSR by showing the environmental sustainability of new businesses and business models [17]. One such example is the "Ugly Fruit" project, reported by Ribeiro *et al.* [26]. In the Ugly Fruit project, products that farmers cannot commercialize through conventional channels are sold to consumers at specific delivery points. On top of the social and economic impacts, LCA proved this measure to be less impactful than the traditional path for these products, which are going to landfills.

CSR might lead to a competitive advantage for companies by making a company gain visibility and a good reputation. Moreover, the repercussion of environmental sustainability assessment techniques, such as LCA, goes beyond helping to achieve environmental sustainability, also leading to greater social sustainability and CSR. As explained by Brehmer *et al.* [12], for instance, the optimal efficiency on the use of materials reduces costs, thus the price of products can be reduced as well, making these products more accessible for the underprivileged population.

3.7. *Strategic planning*

A range of companies have relied on LCA for informed sustainability-related decision-making [21]. From a long-term perspective, LCA has been shown to play a role in helping businesses assess the environmental impacts of investments [27], be it regarding new processes, operations, or business models [28].

The results of LCA applications can be integrated in management dashboards to assist strategic planning and to be more comprehensible for decision makers, allowing to visualize and communicate the results of organizational LCAs as single inventory data or aggregated to impact categories and even in a single environmental sustainability indicator, facilitating decision-making [29, 30].

3.8. *Reverse logistics*

Reverse logistics is directly related to the correct disposal of waste, and for that, LCA can play a major role in guiding the management of the end of life (EOL) of products as well as the return of defective products. A number of problems related to reverse logistics might still be faced in some locations, such as lack of official records (to enable tracking of products at EOL), deficient performance of the public sector with regard to implementing regulations, impact generators not adopting a role of stewardship (polluter pays principle), and, potentially, others [31].

3.9. *Supply chain management*

Another key aspect of sustainable businesses is supply chain management. In many companies, major portions of the environmental impacts of their products are associated with their supply chain [1] and/or the business ecosystem they are part of [32]. Societal and environmental pressures have required organizations to consider sustainability issues across the entire life cycle of products, considering from the conduct of upstream suppliers to the disposal of obsolete products [33]. Therefore, considering a life cycle perspective means that not only the product's input and output flows are accounted for, but the whole supply chain should also be considered [34], and acquiring knowledge on the entire chain of supply can even be a source of value creation for the organization [35].

Nevertheless, sustainable business models can completely modify the entire supply chain of products due to the inclusion of sustainable innovations. In a sharing economy, for instance, peers become suppliers, and this practice increases efficiency in product use and reduces the demand for new products (from virgin materials) [12]. Such advantages can be gauged by means of LCA studies.

3.10. *Legislation and policy*

LCA can contribute to finding gaps in environmental performance (e.g.) of eco-innovative products [10] and, in this sense, it helps achieve and monitor the performance of regulatory requirements, on top of helping to develop these very requirements. When it comes to policy, LCA can be said to be able to guide the allocation of public resources for financing, helping assess the environmental criticality and uncertainty of investments in the search for greener businesses [36].

4. Challenges, Opportunities, and Managerial Implications

A plethora of implications on the use of LCA for sustainable businesses can be observed both in the existing theory and practice. The following paragraphs provide a synthesis of the key challenges, opportunities, and managerial implications as takeaway lessons from this chapter.

Including life cycle thinking in business processes is highlighted as the challenge of the future for sustainable businesses [37]. In this sense, on the one hand, a few challenges can be reported on the use of LCA information when seeking greater sustainability in businesses and organizations. Companies will face challenges and barriers to be overcome within all of the 10 aspects presented in this chapter. Many companies have numerous processes, along with their (many times complex) material and information flows, which require a very well-aligned chain of management, with knowledgeable employees who are able to track and improve them continuously. Accounting for all the phases of the life cycle of a system and seeking innovation to improve its environmental sustainability requires systematic and open-minded thinking, which needs to be deployed on a daily basis. A few business areas might be more impacted than others by LCA results, this will depend on a hierarchical perspective of the organization and, at the same time, on the organizational culture.

On the other hand, opportunities are also present within those aspects in favor of companies that practice life cycle thinking and LCA. Life cycle management can support the environmental, economic, and social aspects that may be linked to sustainable actions. Understanding, working with, and taking advantage of the 10 aspects (or as many as possible) presented in this chapter and being able to actually apply the lessons learned with them in different contexts within the industrial sector can be key in a company's progress. Therefore, companies need to be prepared for barriers and opportunities if they are to remain in the market.

Moreover, a range of managerial implications could be mentioned on the influence of LCA toward more sustainable businesses. One of the outlining issues is the vision an LCA study can provide of a system or a company, a vision that is both holistic (due to all the process mapping) and detailed (due to determining every input and output of each process) at the same time, thus enabling decision makers, business managers, and owners to make informed decisions. Nevertheless, all this knowledge does not come without a cost. LCAs can be quite time- and resource-intensive, needing reasonable time windows to be allotted to LCA projects, on top

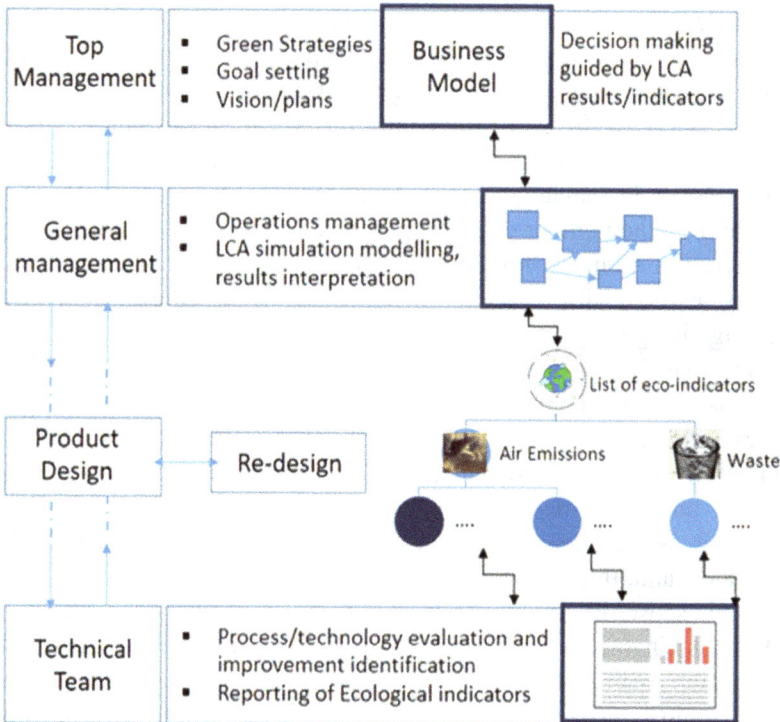

Figure 3.   Framework for green business decision-making driven by LCA indicators.

of needing a specialized workforce highly knowledgeable in (e.g.) process mapping, material and flow analysis, and impact assessment [28, 32, 36]. Overall, LCA results can help guide green business activities and their decision-making procedures move toward more appropriate pathways to ensure more sustainable conduct. A proposed framework depicting the potential inter-connection and interactions between top and general management, product design and re-design divisions, and technical teams using LCA to aid in decision-making for a green business model is illustrated in Figure 3.

5. Concluding Remarks

This chapter aimed to present the contributions of LCA to key business aspects that help direct organizations toward more sustainable business conduct by means of a review of the relevant existing literature.

LCA contributes positively to 10 key aspects that lead toward more sustainable businesses, namely research and development and innovation, product development, operations management, economy, marketing and environmental labeling, corporate social responsibility, strategic planning, reverse logistics, supply chain management, and legislation and policy.

LCA is a powerful tool which helps manage businesses more sustainably and aids in informed sustainable decision-making. However, using LCA to assist in managing a few business aspects is still a challenge depending on the number of areas, processes, inputs, and outputs companies have. Furthermore, for the use of LCA results in business management to be successful, sustainability must be at the core of companies, including changes in the organizational culture and values.

Moreover, understanding the benefits of LCA that might lead toward more sustainable businesses is important because it allows business owners and managers to make informed decisions and to invest more in sustainability. At the same time, it allows businesses to optimize their processes and use of resources, reduce their emissions and wastes, better train their workforce, achieve a better reputation, gain competitive advantages, and, thus, to be more resilient.

In this sense, it is highly suggested to continue analyzing the role of LCA in sustainable businesses as well as of other life cycle thinking techniques, such as life cycle costing and social LCA, in order to achieve a holistic contribution of sustainability (including economic and social dimensions) to business aspects.

Acknowledgments

Authors Murillo Vetroni Barros, Rodrigo Salvador, and Romulo Henrique Gomes de Jesus have received research grants from the Coordenação de Aperfeiçoamento de Pessoal de Nível Superior — Brasil (CAPES) — Finance Code 001. Authors Antonio Carlos de Francisco and Cassiano Moro Piekarski have received research grants from the Conselho Nacional de Desenvolvimento Científico e Tecnológico (CNPq) (sponsored by CNPq 310686/2017-2 and 312285/2019-1).

References

[1] L. L. Kjaer, N. K. Host-Madsen, J. H. Schmidt, and T. C. McAloone, Application of environmental input-output analysis for corporate and product environmental footprints-learnings from three cases. *Sustainability* 7(9), 11438–11461 (2015). doi: 10.3390/su70911438.

[2] X. Li, J. Cao, Z. Liu, and X. Luo, Sustainable business model based on digital twin platform network: The inspiration from Haier's case study in China. *Sustainability* 12(3), 1–26 (2020). doi: 10.3390/su12030936.

[3] N. Klaassen, A. Scheepens, B. Flipsen, and J. Vogtlander, Eco-efficient value creation of residential street lighting systems by simultaneously analysing the value, the costs and the eco-costs during the design and engineering phase. *Energies* 13(13), 3351 (2020). doi: 10.3390/en13133351.

[4] B. Steen, Environmental costs and benefits in life cycle costing. *Manag. Environ. Qual. Int. J.* 16(2), 107–118 (2005). doi: 10.1108/14777830510583128.

[5] C. M. Piekarski, L. M. da Luz, L. Zocche, and A. C. De Francisco, Life cycle assessment as entrepreneurial tool for business management and green innovations. *J. Technol. Manag. Innovation* 8(1), 44–53 (2013). doi: 10.4067/S0718-27242013000100005.

[6] ISO (International Organization for Standardization), *Environmental Management — Life Cycle Assessment — Principles and Framework*, 2nd ed., ISO 14040:2006 (ISO, Geneva, Switzerland, 2006a).

[7] ISO (International Organization for Standardization), *Environmental Management — Life Cycle Assessment — Requirements and Guidelines*, 1st ed., ISO 14044:2006 (ISO, Geneva, Switzerland, 2006b).

[8] R. Salvador, A. C. de Francisco, C. M. Piekarski, and L. M. da Luz, Life Cycle Assessment (LCA) as a tool for business strategy. *Independent J. Manag. Prod.* 5(3), 733–751 (2014). doi: 10.14807/ijmp.v5i3.186.

[9] A. Reid and M. Miedzinski, Eco-innovation: Final report for sectoral innovation watch. Systematic Eco-Innovation Report (Technopolis Group, Brussels, Belgium, 2008). doi: 10.13140/RG.2.1.1748.0089.

[10] R. Barbieri and D. F. L. Santos, Sustainable business models and eco-innovation: A life cycle assessment. *J. Clean. Prod.* **266**, 121954 (2020). doi: 10.1016/j.jclepro.2020.121954.

[11] J. Larsson, Digital innovation for sustainable apparel systems experiences based on projects in textile value chain development. *Res. J. Text. Apparel* **22**(4), 370–389 (2018). doi: 10.1108/rjta-02-2018-0016.

[12] M. Brehmer, K. Podoynitsyna, and F. Langerak, Sustainable business models as boundary-spanning systems of value transfers. *J. Clean. Prod.* **172**, 4514–4531 (2018). doi: 10.1016/j.jclepro.2017.11.083.

[13] R. Travessini, L. Zocche, L. M. da Luz, A. C. Francisco, and A. Braghini, Environmental assessment: Focus of life cycle assessment in the PDP. *Espacios* **35**(2), 20 (2014).

[14] W. Zhang, J. Guo, F. Gu, and X. Gu, Coupling life cycle assessment and life cycle costing as an evaluation tool for developing product service system of high energy-consuming equipment. *J. Clean. Prod.* **183**, 1043–1053 (2018). doi: 10.1016/j.jclepro.2018.02.146.

[15] C. Jiménez-González, A. D. Curzons, D. J. Constable, and V. L. Cunningham, Expanding GSK's solvent selection guide — Application of life cycle assessment to enhance solvent selections. *Clean Technol. Environ. Policy* **7**(1), 42 (2004). doi: 10.1007/s10098-004-0245-z.

[16] S. Roy, Applying Life Cycle Assessment (LCA) in process industry — The chemours experience. In: M. A. Abraham (Ed.), *Encyclopedia of Sustainable Technologies* (Elsevier, Oxford, 2017), pp. 357–362.

[17] J. Singh and T. Cooper, Towards a sustainable business model for plastic shopping bag management in Sweden. *Procedia CIRP* **61**, 679–684 (2017). doi: 10.1016/j.procir.2016.11.268.

[18] C. Butz, J. Liechti, J. Bodin, and S. E. Cornell, Towards defining an environmental investment universe within planetary boundaries. *Sustainability Sci.* **13**, 1031–1044 (2018). doi: 10.1007/s11625-018-0574-1.

[19] F. Mendoza, F. D'Aponte, D. Gualtieri, and A. Azapagic, Disposable baby diapers: Life cycle costs, eco-efficiency and circular economy. *J. Clean. Prod.* **211**, 455–467 (2019). doi: 10.1016/j.jclepro.2018.11.146.

[20] S. Mann and C. Gazzarin, Sustainability indicators for Swiss dairy farms and the general implications for business/government interdependencies. *Int. Rev. Administrative Sci.* **70**(1), 111–121 (2004). doi: 10.1177/0020852304041234.

[21] S. Sim, H. King, and E. Price, *The Role of Science in Shaping Sustainable Business: Unilever Case Study Taking Stock of Industrial Ecology* (Springer International Publishing, 2016), pp. 291–302. https://link.springer.com/chapter/10.1007/978-3-319-20571-7_15

[22] ISO (International Organization for Standardization), ISO 26000: 2010 guidance on social responsibility (2010).

[23] L. Zu, ISO 26000. In: S. O. Idowu, N. Capaldi, L. Zu, and A. D. Gupta (Eds.), *Encyclopedia of Corporate Social Responsibility* (Springer, Berlin, Heidelberg, 2013). doi: 10.1007/978-3-642-28036-8_251.

[24] G. Cardeal, K. Höse, I. Ribeiro, and U. Götze, Sustainable business models–canvas for sustainability, evaluation method, and their application to additive manufacturing in aircraft maintenance. *Sustainability* **12**(21), 9130 (2020). doi: 10.3390/su12219130.

[25] T. Haavaldsen, O. Lædre, G. H. Volden, and J. Lohne, On the concept of sustainability–assessing the sustainability of large public infrastructure investment projects. *Int. J. Sustainable Eng.* **7**(1), 2–12 (2014). doi: 10.1080/19397038.2013.811557.

[26] I. Ribeiro, P. Sobral, P. Pecas, and E. Henriques, A sustainable business model to fight food waste. *J. Clean. Prod.* **177**, 262–275 (2018). doi: 10.1016/j.jclepro.2017.12.200.

[27] H. Zhang, R. Haapala, E. Vanlue, and H. Funk, Environmental impact and cost assessment of product service systems using IDEFO modeling. *Paper presented at the 39th Annual North American Manufacturing Research Conference*, NAMRC39, Corvallis, OR (2011).

[28] E. Scheepens, G. Vogtlander, and C. Brezet, Two life cycle assessment (LCA) based methods to analyse and design complex (regional) circular economy systems. Case: Making water tourism more sustainable. *J. Clean. Prod.* **114**, 257–268 (2016). doi: 10.1016/j.jclepro.2015.05.075.

[29] V. Büdel, A. Fritsch, and A. Oberweis, Integrating sustainability into day-to-day business: A tactical management dashboard for O-LCA. In: *Proceedings of the 7th International Conference on ICT for Sustainability,* pp. 56–65, (2020). doi: 10.1145/3401335.3401665.

[30] J. C. Mann, M. L. Abramczyk, M. R. Andrews, J. A. Rothbart, M. Small, and R. Bailey, Sustainability at Kluge Estate vineyard and winery. *Paper Presented at the 2010 IEEE Systems and Information Engineering Design Symposium*, SIEDS10, Charlottesville, VA (2010).

[31] S. Garcia, I. A. Nääs, C. Neto, and M. dos Reis, Reverse logistics and waste in the textile and clothing production chain in Brazil. In *IFIP International Conference on Advances in Production Management Systems* (Springer, Cham, 2019). pp. 173–179. doi: 10.1007/978-3-030-30000-5_23.

[32] F. Boons and N. Bocken, Assessing the sharing economy: Analyzing ecologies of business models. In: *Proceedings of the PLATE (Product Lifetimes and the Environment) Conference,* Netherlands, pp. 46–50 (2017). doi: 10.3233/978-1-61499-820-4-46.

[33] J. Fiksel, Meeting the challenge of sustainable supply chain management. In *Treatise on Sustainability Science and Engineering* (Springer, Dordrecht, 2013), pp. 269–289. doi: 10.1007/978-94-007-6229-9_16.

[34]   A. Fritsch, Towards a modeling method for business process oriented organizational life cycle assessment. In *Proceedings of the 7th International Conference on ICT for Sustainability,* pp. 200–203 (2020). doi: 10.1145/3401335.3401360.

[35]   V. Julianelli, G. Caiado, L. F. Scavarda, and F. Cruz, Interplay between reverse logistics and circular economy: Critical success factors-based taxonomy and framework. *Resour. Conserv. Recycl.* **158**, 104784 (2020). doi: 10.1016/j.resconrec.2020.104784.

[36]   S. Lokke, H. Schmidt, I. Lyhne, L. Kornov, and R. Revsbeck, How green are supported "green" business models? Time for the life cycle approach to enter public support programmes. *Int. J. Life Cycle Assess.* **25**(10), 2086–2092 (2020). doi: 10.1007/s11367-020-01806-9.

[37]   S. Junnila, Life cycle management of energy-consuming products in companies using IO-LCA. *Int. J. Life Cycle Assess.* **13**(5), 432 (2008). doi: 10.1007/s11367-008-0015-yC.

Chapter 12

Intertwining Ecosystem Services with Life Cycle Assessment: Recommendation for Paradigm Shift

Benedetto RUGANI[*,§]**, Javier Babí ALMENAR**[*,¶]**,
Thomas ELLIOT**[†,‖] **and Benoit OTHONIEL**[‡,**]

*Environmental Research & Innovation (ERIN) Department,
Luxembourg Institute of Science and Technology (LIST) —
41 Rue du Brill, 4422 Belvaux, Luxembourg*

†*Department of Construction Engineering, École de Technologie
Supérieure (ÉTS) — 1100 Notre-Dame O., Montréal,
QC H3C 1K3, Canada*

‡*INRAE, UR ETBX — 50 avenue de Verdun Gazinet,
F-33612 Cestas cedex, France*

§*benedetto.rugani@list.lu*

¶*j.baalm@gmail.com*

‖*thomas.elliot@etsmtl.ca*

**ben.othoniel@gmail.com*

The last decade of scientific literature testifies an increasing interest in including ecosystem services (ES) accounting in life cycle assessment (LCA). Existing inventory models and impact assessment methods intertwining ES with LCA, however, display some methodological caveats such as limited coverage of the analyzed ES flows as well as a fragmented use of integrated frameworks in conventional LCA practices. This chapter provides an overview of the approaches found in LCA to account for ES. It also attempts to extend the current life cycle impact assessment paradigm, customized for the evaluation of detrimental impacts on ES, with an ES assessment approach that considers their beneficial contribution to the ecosystem and human well-being. Accordingly, nature-based solutions applied in urban contexts show unprecedented opportunities to test and foster the incorporation of ES knowledge both at the inventory and impact assessment stages in terms of positive externalities. It is envisaged that future ES assessments will be able to include applicable information concerning both supply and demand of ES, so that future LCA practices can become a full-fledged decision support tool for sustainable management of natural capital.

1. Introduction

Human well-being depends upon numerous ecological goods and services from nature. Sustainable management of natural capital is thus necessary to maintain livable conditions of clean air and water, to keep nutrients regulation cycles, to secure renewable energy and food, and in general, all the biotic and abiotic resources required to guarantee human life on Earth. Such sustainable management, however, is achievable only if the complex cause–effect relationships among the anthroposphere processes and natural cycles are sufficiently comprehended.

One scientifically valuable and widely acknowledged solution to reach this goal is applying a life cycle assessment (LCA) approach to human-driven production systems. With the inception of the ISO 14040–14044 international standards, LCA has gained worldwide recognition for its use in solving environmental sustainability issues [1]. LCA can support private and public organizations in reducing their pressure and possible damage to ecosystems and human health. The rationale of LCA consists of first inventorying all the interactions with the environment caused by goods and services along their life cycle, i.e. from the extraction of raw materials to the end-of-life phases, then assessing the impact that these interactions can generate on biodiversity, natural resource availability, and human health (see further in Section 3).

Researches over the last 30 years have yielded powerful and sophisticated tools to model in detail these inventory and impact assessment phases, improving the overall credibility and robustness of the methodology [2]. Beyond LCA, numerous methods follow a life cycle thinking (LCT) rationale [3], such as environmentally extended input–output frameworks [4] and material flow analysis [5]. All these hybrid environmental accounting methodologies have slightly different but often complementary features. Their use in isolation or in combination allows addressing several methodological questions, such as more and more detailed spatially explicit assessments, high-resolution technology data collection and manipulation, and complex life cycle network modeling.

Despite these achievements, challenges to achieve sustainable management practices of the natural capital by means of LCT approaches still exist. For example, conventional LCA models are unable to adequately evaluate the dynamics of territorial systems that are multifunctional, especially in the area of establishing clear system boundaries and functional units (refer to Chapter 1). Moreover, commonly applied LCA models assume steady-state conditions, which might be inadequate for acquiring non-linear information involved in the sustainable management of local natural capital assets. Finally, there is still a general lack of comprehensive integration of ecosystem services (ES) accounting in LCA [6, 7].

In this regard, the LCA community is working on the creation of a Global Life Cycle Impact Assessment Method (GLAM), which is currently at its third phase [8]. The GLAM implementation initiative, which involves LCA practitioners and method developers in several task forces, aims to identify the "current best available practices" in a variety of areas of relevance for LCA development and innovation. This is done in order to strengthen the synergies among different disciplines, reduce methodological inconsistencies, and ultimately build effective sustainable management tools based on LCA. In recent years, several GLAM working groups were created to discuss ES as a priority issue for impact assessment in LCA [9, 10]. Progress in this direction was crucial to identify existing limitations and areas of further development, with the general ambition to determine effective strategies to assess impacts, either beneficial or detrimental, on the provision of ES associated with the life cycle of production systems.

In order to offer a global picture of the current work in progress, this chapter presents the state of the art on the integration of ES accounting in LCA (Section 2). It further illustrates and discusses new directions of research that may be relevant to enhance the assessment of ES according to an LCT rationale (Section 3). These are based on recent works that developed integrated models combining ES valuation and life cycle-based approaches, both for studying urban nature-based solutions (Section 4) and for applying LCT to urban areas (Section 5).

2. State of the Art

Parallel to the development of advanced LCA methodologies, research in the field of ES has enormously grown over the last 20 years [11]. ES can be defined as "the ecological characteristics, functions, or processes that directly or indirectly contribute to human wellbeing: that is, the benefits that people derive from functioning ecosystems" [12]. To some extent, ES can be interpreted as the outputs of the natural capital, which Guerry *et al.* define as "the living and non-living components of ecosystems — other than people and what they manufacture — that contribute to the generation of goods and services of value for people" [13]. According to CICES [14], which is one of the most recent ES classification systems, three main categories of ES exist: provisioning services (e.g. food, water, bioresources), maintenance & regulation services (e.g. air purification, climate regulation, pollination) and cultural services (i.e. non-material ecosystem outputs that have symbolic, cultural, or intellectual significance). In this sense, ES are produced in all types of ecosystems, from the intensively managed ones (e.g. agroecosystems) to those with low human imprint [13].

The release of Millennium Ecosystem Assessment in the early 2000s [15] represented the first global milestone for the harmonization of ES-related concepts and indicators into an internationally acknowledged classification system. Such evidence fostered the LCA scientific community toward the combination between LCA and the assessment of ES approximately 10 years ago [16]. Over the last decade, LCA scholars have explored several ways to integrate ES in LCA, finding clear evidence about the share of environmental sustainability objectives between the two research areas [6, 7]. A recent systematic critical review of the literature highlights that approximately 19 LCA studies have attempted to

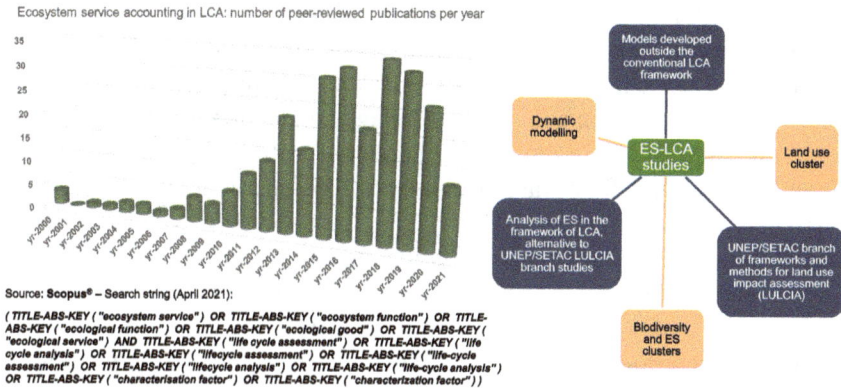

Figure 1. Evolution of studies integrating or accounting for ES in (models of) LCA. Data collection from Scopus® performed on April 2021.

integrate biotic ES in various meaningful ways [7]. As reflected in this and other previous literature analyses [6, 17, 18], three main schools of thought essentially took hold so far, which include almost 300 peer-reviewed studies, the majority of which were published in 2010 (see Figure 1). Rugani *et al.* [6] regrouped them into three corresponding lines of research: (1) "UNEP/SETAC branch of frameworks and methods for land-use impact assessment (LULCIA)", (2) "Analysis of ES in the framework of LCA, alternative to UNEP/SETAC LULCIA branch studies", and (3) "Models developed outside the conventional LCA framework". Likewise, VanderWilde and Newell [7] clustered the LCA-ES literature into three main sets called "Biodiversity and ES clusters", "Land-use cluster", and "Dynamic modeling". Despite the minor differences between those two categorizations, many overlaps occur that may result in similar findings and conclusions.

In detail, a first and widespread effort in merging ES into LCA was performed within the UNEP/SETAC LULCIA initiative, starting from the evaluation scope of LCA, which aimed to identify and characterize detrimental impacts generated by human processes (driven by, e.g. land use) on the provision of ES. This brought to the first development of characterization factors (CFs) for some ES [19]. The UNEP/SETAC LULCIA initiative was therefore specifically oriented toward developing CFs for land-use impacts on biodiversity and ES in the Life Cycle Impact Assessment (LCIA) [19–21]. A common rationale of all the UNEP/

SETAC LULCIA studies was to quantify and deliver CFs compatible with existing life cycle inventory (LCI) datasets including information on land use and land-use change. These efforts led to the implementation of CFs for assessing impacts on a few provisioning and maintenance & regulation services, namely biotic production potential [22], climate regulation potential [23], and freshwater regulation, erosion regulation and water purification potentials [24]. These CFs were developed for midpoint impact assessment using physical units and have been conveyed into an ES damage potential at the endpoint level. Further research efforts proceeded to designate endpoint CFs in monetary terms [25]. Indicators of impacts on functional diversity [26] and species richness [27] were also implemented with the aim to translate direct impacts on species habitats into a biodiversity damage potential. While CFs developed within and in accordance with the LULCIA initiative still represent the most compatible and "ready-to-use" values to perform ES-LCA analyses, their implementation in LCA practice is very limited. One reason might be their inability to link ES with the final value and actual benefits to human society.

These methodological limitations led to the second effort of the LCA community to account for ES in LCA [28]. The goal was to use integrated modeling frameworks implemented outside the LCA field as impact characterization models for ES in order to capture complex ecological dynamics in the CFs. More specifically, such research included methodologies, applications, and tools developed *ad hoc* to assess the impact of life cycle activities on the provision of specific ES. Different models and impact drivers were implemented and considered, but they were usually not comparable among each other, as they were not harmonized under an umbrella approach like the LULCIA [29–35]. A commonality of these studies was to quantify CFs for possible application in existing LCIA frameworks or to develop new impact characterization solutions that match well with the current LCA calculation structure. These approaches could bring more sophistication and diversity for the following: (i) in the calculation of CFs (e.g. higher spatial resolution and/or time dependency and/or wider and deeper geographical distribution) [8] and (ii) in the choice of the impact drivers (not only land use was considered but also other stressors like the extraction of natural resources [6]). Despite these improvements, as for the previous group of studies, the application of one or another of those models remains very limited and case-specific, most probably because of the lack of harmonization into one or another methodological solution.

Finally, a few proposals of LCA-ES integration were developed outside the typical LCIA framework [36–38]. All of them made use of the existing ES modeling and assessment tools in order to account for damages on ES applying specific life cycle data or inventory models or simply adopting an LCT without developing and using CFs. Some insights from those novel computational frameworks for LCA-ES integration, which make use of, e.g. integrated system dynamics-based models or other deterministic tools [6], can inspire the future development of a new generation of impact characterization models to assess ES in LCA. Further lines of reasoning are given in Section 3, which focuses on the limitations and opportunities to implement the third generation of approaches to assess ES following an LCT rationale.

3. Deepening into the ES-LCA Modeling Paradigm

LCA typically aims to evaluate the impact on some indicators of interest, which in our case would represent the provision of ES over the life cycle of a selected functional unit (FU). As an example of FU, a common textile material like a pair of socks is considered here. The supply–production stages associated with the FU may include the growing and picking of cotton, sewing and conditioning of textile materials, packaging, transportation and distribution, its daily use and finally its incineration or other end-of-life options. Beyond these directly involved processes, which are together called the "foreground system" of the FU, many other processes (e.g. production and use of energy upstream, logistic arrangements) belong to the supply–production chain indirectly, constituting the "background system". For example, energy has to be produced to operate these processes, agriculture activities involved in cultivating cotton needs irrigation, use of fertilizers, etc. In most cases, and as it is assumed here, LCA practitioners have control in data collection and modeling for the foreground system and not for the background system.

Hence, LCA starts from what can be described as the vector x of size N, with N being the number of all possible processes in the world, such that $x_i = a$ means that an amount a of products from the foreground process i is needed to obtain the final sock. Note that with a few processes in the foreground system, most of the values in x are zeroes. Once x is defined, LCA is applied in two steps.

The first step consists of calculating the LCI of the FU. This yields two types of results, that is, the needed amounts of products from all the background and foreground processes and the amounts of environmental stressors belonging to these processes (refer again to Chapter 1). The former can be defined as x^*, as similar to x but with less non-zero values and the latter as the vector s of size M, with M being the number of possible stressors, such that $s_i = b$ means that a total amount b of the stressor i is due to the life cycle of the sock. Note that the stressor i can come from several processes like, for instance, the CO_2 emitted from transportation and energy production processes. In this sense, the range of accounted stressors is wide. It includes the release of (eco)toxic substances, the emission of greenhouse gases, the use of natural resources, etc.

The second step, called LCIA, consists of assessing the impact of the inventoried stressors on the indicators of interest. To do so, pre-calculated CFs are used, as anticipated in Section 2. However, such an operation does raise the following issues:

- The location where the stressors occur is unknown or not precise (at best at the national scale), while ES significantly vary over small distances.
- The provision of ES is not constant over time; the time of occurrence and the duration of the stressors can significantly affect the final change in ES whereby LCI and CFs are not time-dependent.
- The total impact on ES of multiple stressors is not equal to the sum of their individual impacts; however, by design, CFs associate a single stressor to a single ES or multiple ES at best, but not several stressors to single or multiple ES.
- ES provision is non-linear with respect to its conditions, but the models underlying LCI and LCIA are linear, as a consequence, the impacts on ES that can be characterized with LCA may be inaccurate because of the different life cycle spatial and temporal scales.

To address these issues, some solutions can be identified as anticipated in Section 2, although not free of limitations. A further regionalization of the LCI and, consequently, of the CFs might be foreseen. However, regionalizing the background system to an appropriate scale representative of the modeled supply chain is impractical due to the huge complexity and variability of the network of processes involved. Improving the dynamic modeling of the LCA model — with the aim to build time-dependent LCIs and calculate time-dependent CFs — could be beneficial. But so far, no operational solutions are available to develop such time-dependent LCIs for the background system [39]. Furthermore, instead of using CFs, models that take as inputs multiple stressors could be used to perform the LCIA stage and assess impacts on ES. However, an issue occurs that this will not be scalable (cf. the following). At last, non-linear models could be used to simulate the LCI, performing a non-linear accounting for each process with economic modeling approaches (e.g. general equilibrium models) to determine which process comes from where. Such an integrated model may further include ES models to perform the whole LCA using a single calculation framework [35]. Nevertheless, this would be difficult to conceive as it would require substantial sophistication in data collection and manipulation along with an increase in computational power. Hence, the use of such modeling techniques for daily LCA investigations still needs to reach a state of consensus based on proven operability. Given these limitations, an intermediate solution based on the following criteria seems to be a reasonable choice:

- *Focus the efforts on improving the assessment of ES on the foreground system*: Since this is usually known (the practitioner knows which industry, where it is located, what is involved in), applying existing ES models, which may take multiple stressors as inputs and are at a high spatial resolution, seems feasible.
- *Improve the uncertainty analysis of CFs by providing their values as ranges instead of single values*: Calculating scenario-dependent CFs, where the scenarios represent possible broad changes of the world (e.g. calculate CFs for different levels of temperature rise) could help solve some issues with interacting stressors (even if not all the possible ones) and, at least provide a better idea of the potential change in ES induced by the FU.
- *Foster the operability of other models to use them simultaneously with LCA and take into account positive impacts on ES*: LCA was originally designed to address impact-oriented questions about the overextraction of natural resources, the overuse of land, or the overemission of pollutants. Hence, it may not be reasonable to address the question of how to incorporate knowledge on the benefits from and the damages to the natural capital and ES, which is revealed to be far more complex than solving the rationale "what if I consume or emit more". Acknowledging this limitation and then promoting research on the inclusion of positive impacts within the existing LCIA modeling approaches thus seems to be a wise compromise.

A proposal for the integration of ES in LCA, both at the level of inventory and impact assessment (with the notion of positive and negative externalities), is illustrated in Section 4, focusing on the concept of "nature-based solutions" (NBS). These are considered to address multiple challenges through the generation of ES bundles [40], thus becoming unprecedented guinea pigs for analyzing the integration of ES in LCA.

4. Nature-Based Solutions for Intertwining LCA and ES

Several ES assessment methods exist to assess positive environmental, economic, and social values derived from low to heavily managed ecosystems [41–43]. Like most LCA methods, ES as environmental values (or impacts) are usually estimated making use of biophysical assessment methods. These measure biophysical attributes (i.e. ecosystem structure) and ecological processes as parameter proxies of ES flow. However, biophysical methods are not capable of offering a complete view of the economic welfare derived from ES. Consequently, once biophysical ES accounts are obtained, the final goods and benefits derived from ES are usually calculated in monetary units. In this sense, monetization of ES accounts acts as a kind of endpoint impact on human well-being. Accounting for them as part of LCA, by considering how changes in ecosystem structures influence ES accounts (midpoint impact), and consequently human well-being (endpoint impact) would support the complete sustainable management of the natural capital.

For LCA studies focused on evaluating environmental impacts derived from changes in territorial systems or ecosystems, the incorporation of ES accounts could have strong relevance in improving the completeness and accuracy of the assessments [35, 36]. However, as anticipated in Section 2, the integration of ES is still far from achieving a state of full operability in LCA. Besides issues with the conceptual integration, the development of impact characterization models for multiple ES that remain valid for multiple geographical contexts and spatial scales is a very complex and data-demanding procedure. As an intermediate compromise, a more feasible approach could be to intertwine ES and LCA like outlined in Figure 2. This intertwining of ES and LCA has been successfully applied to evaluate the impact of NBS implemented in urban areas as part of Nature4Cities (https://www.nature4cities.eu/). NBS are

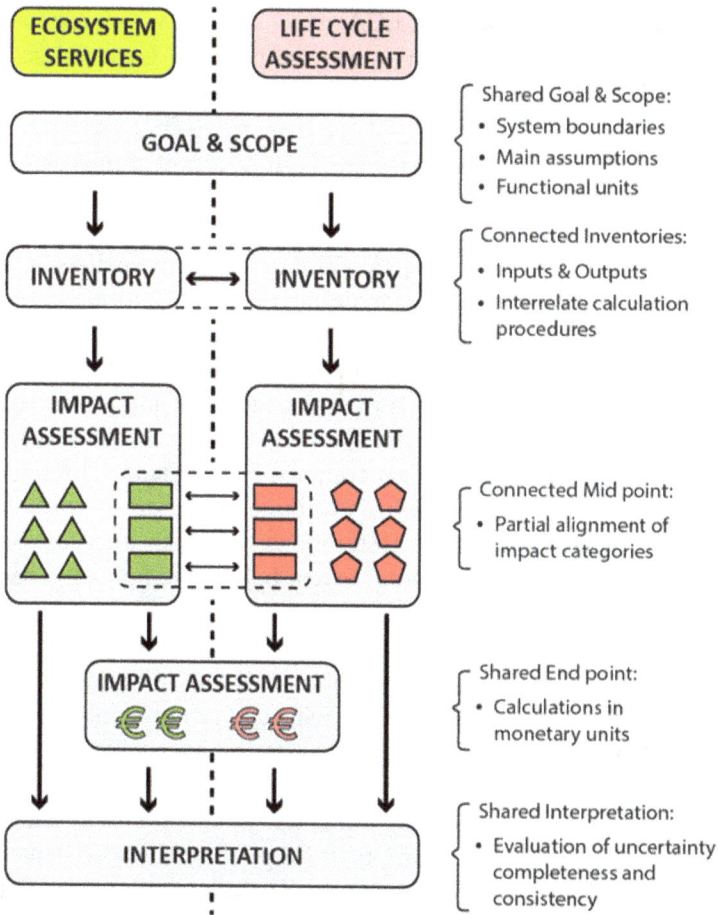

Figure 2.   Conceptual diagram of the intertwining of ES accounting and LCA.

defined as solutions supported by nature that produce environmental, social, and/or economic benefits in a cost-effective way [44]. Like for other ecosystem structures (or actions applied on them), the main way for these benefits to arise is through ES supply. However, human actions applied on NBS (e.g. planting, irrigation) over their life cycle have financial costs and produce negative environmental impacts that should be accounted for [45, 46]. The following paragraphs briefly describe the lessons learned from intertwining ES and LCA, as illustrated in Figure 2, in order to evaluate the impact of urban NBS. The description emphasizes the adjustments required in each LCA stage, most of which are valid for any type of ecosystem structure. When defining the goal and scope of LCA applied to NBS, the following conditions should be considered:

- FUs are spatially defined based on their physical properties or the amount of people or area that get benefits from the supplied ES. For example, for urban forests or green roofs, impacts per square meter are usually used to define an FU [45, 47]. For solutions such as bioretention basins, the functional unit is defined according to the amount of impermeabilized area that the solution serves or the potential amount of water volume that they could receive from them over their operational life [48, 49]. In terms of the lifetime extent, ecosystems do not have a specific end of life, they evolve over time and might persist across several human generations. Consequently, lifetimes are defined based on average lifetimes of building structures with analog function or up to when major restorations or human interventions are expected or keeping the temporal extent below one human generation [45, 50, 51].
- System boundaries definition should acknowledge that modeling of ES flows requires a detailed characterization of the abiotic and biotic structures. Ecosystem structures, heavily managed or not, evolve over time and are dependent on their position in the broader territorial system, weather conditions and people's behaviors. Consequently, their ES performance evolves in the long term (e.g. years) and in the short term (e.g. seasons). Moreover, in part, their ES performance also depends on other surrounding ecosystem structures. Therefore, unless very detailed spatio-temporal data across the supply chain occur, ES accounting should be restricted to the foreground level, where their dynamism should be taken into account.
- For the purpose of defining processes studied in the background and foreground levels, the life cycle of NBS can be organized in the following three phases: implementation (i.e. from sowing or raw material extraction until the implementation of the NBS), operational (i.e. lifetime of the implemented solution), and end-of-life (i.e. from the collection of dead components up to the final waste treatment). In the implementation phase, indirect ES flows cannot be easily accounted for, being part of the background system (Section 3). For the end-of-life phase, ES accounts can be used to feed the LCI with specific inputs, e.g. dead biomass generated on-site that would be treated as waste or

raw material. Moreover, contrary to most goods and services modeled in LCA, for NBS these phases do not have a single start and end in time. For example, an urban forest might have a recurrent implementation of new individual trees over time substituting dead trees. Similarly, the end of life of individual trees or part of their components (e.g. branches) will occur at different points in time. Here, the main element evaluated is the ecosystem structure, not its individual biotic or abiotic components. Consequently, evaluating these independently as if they were products will not fully inform about the ES produced because interactions are relevant, and simple aggregation of outputs per component cannot be applied. In the case of NBS, life cycle phases partially overlap, and the modeling framework used for the evaluation should permit this.

When building the inventories, data about the local geobiosphere and anthroposphere need to be interrelated. The local climatic, abiotic, and biotic conditions in which the NBS will be implemented as well as its size will influence some technosphere processes in the implementation, operational, and end-of-life phases. For example, urban forests planted in excessively compacted soil would require more working effort (manual or through machinery) than non-compacted soils. As another example, a peri-urban forest and a small urban forest in the core of a dense city (i.e. with more human presence) will require management with a different frequency and some of the management actions might be different. Moreover, a small urban forest (<2 ha) compared to large urban forests placed in an equivalent context might supply less ES and require more management per square meter. Hence, interrelating local data about the ecosystem structures and LCIs about technosphere processes allows developing accurate inventories as well as their transposition to equivalent case studies.

In the case of NBS, the impact assessment should be performed at both midpoint and endpoint levels. The selection of impact categories, their indicators, and therefore specific LCA and ES methods offers an opportunity to align LCA and ES frameworks. Like generally accepted in LCA, the estimation of midpoint impact categories involves a lower level of uncertainty than the estimation of endpoint impacts. In addition, and more importantly, midpoint impact indicators represent negative environmental values in the form of environmental effects, being analogs as how environmental values are represented in the form of ES, thus representing impacts at similar levels of abstraction facilitating linking of ES and LCA approaches. Moreover, in LCIA methods with large coverage such as ReCiPe [52], indicators of some midpoint impact categories are analogs to those typically used as a proxy for some ES classes [32]. This would permit to align some categories and quantify the net impacts for them, since equivalent units of measurement are used.

In order to quantify the final net impact on human well-being, monetization of environmental values as externalities will be required. This is due to the current impossibility of aligning all ES classes and impact categories. In particular, the use of monetary values would facilitate the calculation of endpoints and consideration of structural uncertainty. Such monetary values should be valid for a broad geographical context that incorporates low, central, and high monetary values per type of midpoint impact, using environmental pricing methods [53]. As a complement, the quantification of financial values associated with human actions applied on NBS would provide a complete LCA framework, where the endpoint would be assessed with a full environmental life cycle costing or cost–benefit analysis [54, 55].

Finally, the interpretation of an assessment of NBS intertwining ES and LCA would be equivalent to the one of a product life cycle, but the spatial aspects would be described in detail within the limitations, e.g. constraints regarding the relationships of surrounding ecosystems accounted for as part of the modeling, or if it were not possible to take into account relationships with surrounding ecosystems, as part of the ES accounting and inventorying of the expected management actions applied on NBS.

5. Use of Life Cycle Thinking in Urban Ecosystem Services Assessments for Remote Impacts on Demand

Urbanized contexts are optimal for NBS implementations, whereby a general lack of ES (i.e. low supply) and an overload of environmental impacts usually take place. This can even generate relevant displacements elsewhere, increasing the demand for ES to mitigate or address impacts outside the city. Therefore, ES are subject to such a "supply" and "demand" function ([37]; see also the framework in Figure 1), much like other types of economic services. An ES supply refers to the provision of an ES by an ecosystem. While demand for an ES is determined by a community's accessibility to and need for a given ES [56]. When the said ecosystem suffers changes to its health, the supply of its ES is affected and the result is an undersupply of demand.

When conducting urban ES assessments, it is typical to consider both the ES supply and ES demand from ecosystems within the city, which can ideally be represented throughout an urban metabolism model based on LCIs [57]. In cases where the supply and demand exist in a closed system, this approach is fit for this purpose. For instance, "Land Footprint" was considered in a simplified example to account for agricultural areas used to provide bio-resources in a global supply–demand of a bio-derived product [58]. However, cities are dependent on imported resources from remote ecosystems, such as with food, energy, and materials that are subject to complex global supply chains [59, 60]. Acknowledging the dependence on these remote ecosystems to support urban life requires an assessment framework that accounts for the impact urban life has on those remote ecosystems. LCT can be used to better capture the environmental impacts drawn out along multiple resource flows entering the urban system [59, 61]. Accounting for those distal indirect environmental impacts allows for the encapsulation of non-local demand for ES [62], as illustrated in Figure 3. Each resource flow entering the urban system has a life cycle. The resource's constituent parts are extracted, processed, and transported across a variety of ecosystems typically outside the city. After its use phase in the city, waste flows are

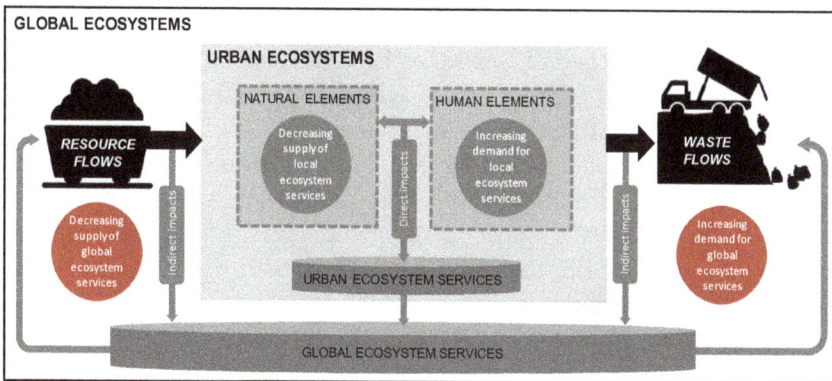

Figure 3.   Relationship between local and global ES using an LCA approach.
Source: Adapted from Ref. [57].

typically exported back to global sinks. Estimating the impact these resources have on global ecosystems before entering the city is achieved by linking each resource to life cycle impact factors, e.g. calculated from LCI databases [62]. These indirect impacts occur in global ecosystems contributing to an undersupply of ES from those ecosystems and thus the demand for ES is generated by the urban supply chain. Assessing the impact these resources have at the end-of-life phase is achieved by linking each resource to the direct environmental impacts. This is relevant, e.g. for combustion activities within the city, for example, fossil fuels used for transport and heating [62]. Depending on the ES being assessed, it may be relevant for emissions to wastewater and soil. In the example of the ES global climate regulation, supply can be measured as carbon storage (e.g. mass of trees sequestering and storing carbon dioxide) and demand measured as global warming potential (e.g. mass of carbon dioxide equivalents emitted by human activities). A city's global climate regulation must then quantify the indirect global warming potential of the resource flows entering the city and the direct global warming potential of the combustion flows exiting the city. The sum of these approximates the additional demand the urban activities place on ecosystems to supply global warming potential.

6. Conclusion and Outlook

This chapter has attempted to synthetize the last decade of research practice and conceptual advancements regarding the ES accounting for LCA. Extending the current LCIA paradigm, customized on the evaluation of the detrimental impacts on ES, with an ES assessment approach that considers their beneficial contribution to well-being, can open methodological avenues to create frameworks integrating impacts in terms of positive externalities. To this end, nature-based solutions applied in urban contexts show first-time opportunities to foster the incorporation of ES knowledge both at the inventory and impact assessment stages. ES assessments can include the supply of ES, the demand of ES, or both. However, the valuation of ES must consider both sides of the coin in order to deploy LCA consistently as a decision support tool for sustainable management of natural capital. To this end, the theory and practice to account for ES in LCA are in their full stages of development, proving more and more sophistication in integrating these two areas of research exists.

References

[1] *ISO 14040:2006: Environmental Management — Life Cycle Assessment — Principles and Framework* (International Organization for Standardization (ISO), Geneva, Switzerland, 2006).

[2] M. Z. Hauschild, R. K. Rosenbaum, and S. I. Olsen, *Life Cycle Assessment — Theory and Practice* (Springer International Publishing AG, Cham, Switzerland, 2018), p. 1216.

[3] G. Sonnemann, E. D. Gemechu, S. Sala, E. M. Schau, K. Allacker, R. Pant *et al.*, Life cycle thinking and the use of LCA in policies around the world. In: M. Z. Hauschild, R. K. Rosenbaum, and S. I. Olsen (Eds.), *Life Cycle Assessment: Theory and Practice* (Springer International Publishing, Cham, 2018), pp. 429–463.

[4] R. H. Crawford, P.-A. Bontinck, A. Stephan, T. Wiedmann, and M. Yu, Hybrid life cycle inventory methods — A review. *J. Clean. Prod.* **172**, 1273–1288 (2018).

[5] D. Laner and H. Rechberger, Material flow analysis. In: M. Finkbeiner (Ed.), *Special Types of Life Cycle Assessment* (Springer Science+Business Media, Dordrecht, NL, 2016). pp. 293–332.

[6] B. Rugani, D. Maia de Souza, B. P. Weidema, J. Bare, B. Bakshi, B. Grann *et al.*, Towards integrating the ecosystem services cascade framework within the Life Cycle Assessment (LCA) cause-effect methodology. *Sci. Tot. Environ.* **690**, 1284–1298 (2019).

[7] C. P. VanderWilde and J. P. Newell, Ecosystem services and life cycle assessment: A bibliometric review. *Resour. Conserv. Recy.* **169**, 105461 (2021).

[8] Life-Cycle-Initiative, *Global LCIA Guidance (GLAM) Phase 3 — "Creation of a Global Life Cycle Impact Assessment Method"* (UN Environment Programme, 2021).

[9] F. Verones, J. Bare, C. Bulle, R. Frischknecht, M. Hauschild, S. Hellweg *et al.*, LCIA framework and cross-cutting issues guidance within the UNEP-SETAC Life Cycle Initiative. *J. Clean. Prod.* **161**(Supplement C), 957–967 (2017).

[10] A. Antón, D. Maia de Souza, F. Teillard, and L. Milà i Canals, Addressing biodiversity and ecosystem services in Life Cycle Assessment. In: D. Geneletti (Ed.), *Handbook on Biodiversity and Ecosystem Services in Impact Assessment* (Edward Elgar Publishing, Inc., Cheltenham, UK, 2016), pp. 140–164.

[11] A. V. Torres, C. Tiwari, and S. F. Atkinson, Progress in ecosystem services research: A guide for scholars and practitioners. *Ecosyst. Serv.* **49**, 101267 (2021).

[12] R. Costanza, R. de Groot, L. Braat, I. Kubiszewski, L. Fioramonti, and P. Sutton *et al.*, Twenty years of ecosystem services: How far have we come and how far do we still need to go? *Ecosyst. Serv.* **28**, 1–16 (2017).

[13] A. D. Guerry, S. Polasky, J. Lubchenco, R. Chaplin-Kramer, G. C. Daily, R. Griffin *et al.*, Natural capital and ecosystem services informing decisions: From promise to practice. *Proc. Nat. Acad. Sci. U.S.A* **112**(24), 7348–7355 (2015).

[14] R. Haines-Young and M. Potschin, *Common International Classification of Ecosystem Services (CICES) V5.1 and Guidance on the Application of the Revised Structure* (Fabis Consulting Ltd, The Paddocks, Chestnut Lane, Barton in Fabis, Nottingham, NG11 0AE, UK, 2018). www.cices.com.

[15] MEA, *Environmental Degradation and Human Well-Being: Report of the Millennium Ecosystem Assessment* (Blackwell Publishing Ltd., Oxford, UK, 2005), pp. 389–398.

[16] Y. I. Zhang, S. Singh, and B. R. Bakshi, Accounting for ecosystem services in life cycle assessment part I: A critical review. *Environ. Sci. Technol.* **44**(7), 2232–2242 (2010).

[17] D. Maia de Souza, G. R. Lopes, J. Hansson, and K. Hansen, Ecosystem services in life cycle assessment: A synthesis of knowledge and recommendations for biofuels. *Ecosyst. Serv.* **30**, 200–210 (2018).

[18] B. Othoniel, B. Rugani, R. Heijungs, E. Benetto, and C. Withagen, Assessment of life cycle impacts on ecosystem services: Promise, problems, and prospects. *Environ. Sci. Technol.* **50**(3), 1077–1092 (2016).

[19] T. Koellner and R. Geyer, Global land use impact assessment on biodiversity and ecosystem services in LCA. *Int. J. Life Cycle Assess.* **18**(6), 1185–1187 (2013).

[20] T. Koellner, L. De Baan, T. Beck, M. Brandão, B. Civit, M. Margni *et al.*, UNEP-SETAC guideline on global land use impact assessment on biodiversity and ecosystem services in LCA. *Int. J. Life Cycle Assess.* **18**(6), 1188–1202 (2013).

[21] R. F. M. Teixeira, D. M. D. Souza, M. P. Curran, A. Antón, O. Michelsen, and L. M. I. Canals, Towards consensus on land use impacts on biodiversity in LCA: UNEP/SETAC Life Cycle initiative preliminary recommendations based on expert contributions. *J. Clean. Prod.* **112**, 4283–4287 (2016).

[22] M. Brandão and L. I Canals, Global characterisation factors to assess land use impacts on biotic production. *Int. J. Life Cycle Assess.* **18**(6), 1243–1252 (2013).

[23] R. Müller-Wenk and M. Brandão, Climatic impact of land use in LCA-carbon transfers between vegetation/soil and air. *Int. J. Life Cycle Assess.* **15**(2), 172–182 (2010).

[24] R. Saad, T. Koellner, and M. Margni, Land use impacts on freshwater regulation, erosion regulation, and water purification: A spatial approach for a global scale level. *Int. J. Life Cycle Assess.* **18**(6), 1253–1264 (2013).

[25] V. Cao, M. Margni, B. D. Favis, and L. Deschênes, Aggregated indicator to assess land use impacts in life cycle assessment (LCA) based on the economic value of ecosystem services. *J. Clean. Prod.* **94**, 56–66 (2015).

[26] D. de Souza, D. B. Flynn, F. DeClerck, R. Rosenbaum, H. de Melo Lisboa, and T. Koellner, Land use impacts on biodiversity in LCA: Proposal of characterization factors based on functional diversity. *Int. J. Life Cycle Assess.* **18**(6), 1231–1242 (2013).

[27] L. de Baan, R. Alkemade, and T. Koellner, Land use impacts on biodiversity in LCA: A global approach. *Int. J. Life Cycle Assess.* **18**(6), 1216–1230 (2013).

[28] D. Arbault, M. Rivière, B. Rugani, E. Benetto, and L. Tiruta-Barna, Integrated earth system dynamic modeling for life cycle impact assessment of ecosystem services. *Sci. Tot. Environ.* **472**, 262–272 (2014).

[29] R. van Zelm, M. van der Velde, J. Balkovic, M. Čengić, P. M. F. Elshout, T. Koellner *et al.*, Spatially explicit life cycle impact assessment for soil erosion from global crop production. *Ecosyst. Serv.* **30**, 220–227 (2018).

[30] T. Schaubroeck, R. A. F. Alvarenga, K. Verheyen, B. Muys, and J. Dewulf, Quantifying the environmental impact of an integrated human/industrial-natural system using life cycle assessment; A case study on a forest and wood processing chain. *Environ. Sci. Technol.* **47**(23), 13578–13586 (2013).

[31] U. Bos, R. Horn, T. Beck, J. P. Lindner, and M. Fischer, *LANCA® Characterization Factors for Life Cycle Impact Assessment — Version 2.0* (Fraunhofer Verlag, Fraunhofer Institute for Building Physics (Germany), Stuttgart, 2016), p. 164.

[32] E. M. Alejandre, P. M. van Bodegom, and J. B. Guinée, Towards an optimal coverage of ecosystem services in LCA. *J. Clean. Prod.* **231**, 714–722 (2019).

[33] H. K. Jeswani, S. Hellweg, and A. Azapagic, Accounting for land use, biodiversity and ecosystem services in life cycle assessment: Impacts of breakfast cereals. *Sci. Tot. Environ.* **645**, 51–59 (2018).

[34] C. Bulle, M. Margni, L. Patouillard, A.-M. Boulay, G. Bourgault, V. De Bruille *et al.*, IMPACT World+: A globally regionalized life cycle impact assessment method. *Int. J. Life Cycle Assess.* **24**(9), 1653–1674 (2019).

[35] B. Othoniel, B. Rugani, R. Heijungs, M. Beyer, M. Machwitz, and P. Post, An improved life cycle impact assessment principle for assessing the impact of land use on ecosystem services. *Sci. Tot. Environ.* **693**, 133374 (2019).

[36] R. Chaplin-Kramer, S. Sim, P. Hamel, B. Bryant, R. Noe, C. Mueller *et al.*, Life cycle assessment needs predictive spatial modelling for biodiversity and ecosystem services. *Nat. Commun.* **8**, 15065 (2017).

[37] X. Liu and B. R. Bakshi, Ecosystem services in life cycle assessment while encouraging techno-ecological synergies. *J. Ind. Ecol.* **23**(2), 347–360 (2019).

[38] X. Liu, B. R. Bakshi, B. Rugani, D. M. de Souza, J. Bare, J. M. Johnston *et al.*, Quantification and valuation of ecosystem services in life cycle

assessment: Application of the cascade framework to rice farming systems. *Sci. Tot. Environ.* **747**, 141278 (2020).

[39]   Y. Pigné, T. N. Gutiérrez, T. Gibon, T. Schaubroeck, E. Popovici, A. H. Shimako *et al.*, A tool to operationalize dynamic LCA, including time differentiation on the complete background database. *Int. J. Life Cycle Assess.* **25**(2), 267–279 (2020).

[40]   J. Babí Almenar, T. Elliot, B. Rugani, P. Bodénan, T. Navarrete Gutierrez, G. Sonnemann *et al.*, Nexus between nature-based solutions, ecosystem services and urban challenges. *Land Use Pol.* **100**, 104898 (2021).

[41]   X. Cheng, S. Van Damme, L. Li, and P. Uyttenhove, Evaluation of cultural ecosystem services: A review of methods. *Ecosyst. Serv.* **37**, 100925 (2019).

[42]   K. G. Turner, S. Anderson, M. Gonzales-Chang, R. Costanza, S. Courville, T. Dalgaard *et al.*, A review of methods, data, and models to assess changes in the value of ecosystem services from land degradation and restoration. *Ecol. Modell.* **319**, 190–207 (2016).

[43]   F. Santos-Martin, A. Viinikka, L. Mononen, L. Brander, P. Vihervaara, I. Liekens *et al.*, Creating an operational database for Ecosystems Services Mapping and Assessment Methods. *One Ecosys.* **3**, e26719 (2018).

[44]   European-Commission, *Towards an EU Research and Innovation Policy Agenda for Nature-Based Solutions & Re-Naturing Cities — Final Report of the Horizon 2020 Expert Group on "Nature-Based Solutions and Re-Naturing Cities"*, DG Research and Innovation 2015, E. Directorate-General for Research and Innovation 2015 Climate Action, Resource Efficiency and Raw Materials EN (full version), European Commission, Brussels (2015).

[45]   E. G. McPherson, A. Kendall, and S. Albers, Life cycle assessment of carbon dioxide for different arboricultural practices in Los Angeles, CA. *Urban For. Urban Green.* **14**(2), 388–397 (2015).

[46]   A. C. Petri, A. K. Koeser, S. T. Lovell, and D. Ingram, How green are trees? — Using life cycle assessment methods to assess net environmental benefits. *J. Environ. Hortic.* **34**(4), 101–110 (2016).

[47]   P. Vacek, K. Struhala, and L. Matějka, Life-cycle study on semi intensive green roofs. *J. Clean. Prod.* **154**, 203–213 (2017).

[48]   C. Xu, J. Hong, H. Jia, S. Liang, and T. Xu, Life cycle environmental and economic assessment of a LID-BMP treatment train system: A case study in China. *J. Clean. Prod.* **149**, 227–237 (2017).

[49]   D. Vineyard, W. W. Ingwersen, T. R. Hawkins, X. Xue, B. Demeke, and W. Shuster, Comparing green and grey infrastructure using life cycle cost and environmental impact: A rain garden case study in Cincinnati, OH. *J. Am. Water. Resour. Assoc.* **51**(5), 1342–1360 (2015).

[50]   P. Rosasco and K. Perini, Evaluating the economic sustainability of a vertical greening system: A cost-benefit analysis of a pilot project in mediterranean area. *Build. Environ.* **142**, 524–533 (2018).

[51] J. Sproul, M. P. Wan, B. H. Mandel, and A. H. Rosenfeld, Economic comparison of white, green, and black flat roofs in the United States. *Energ. Buildings* **71**, 20–27 (2014).

[52] M. A. J. Huijbregts, Z. J. N. Steinmann, P. M. F. Elshout, G. Stam, F. Verones, M. Vieira *et al.*, *ReCiPe2016 — A harmonized life cycle impact assessment method at midpoint and endpoint level* (Department of Environmental Science, Radboud University Nijmegen, 2016).

[53] S. De Bruyn, S. Ahdour, M. Bijleveld, L. De Graaff, E. Schep, A. Schroten *et al.*, *Environmental Prices Handbook 2017 – Methods and Numbers for Valuation of Environmental Impacts* (CE Delft, Delft, 2018), p. 176.

[54] R. Hoogmartens, S. Van Passel, K. Van Acker, and M. Dubois, Bridging the gap between LCA, LCC and CBA as sustainability assessment tools. *Environ. Impact Assess. Rev.* **48**, 27–33 (2014).

[55] T. Schaubroeck, C. Petucco, and E. Benetto, Evaluate impact also per stakeholder in sustainability assessment, especially for financial analysis of circular economy initiatives. *Resour. Conserv. Recycl.* **150**, 104411 (2019).

[56] R.-U. Syrbe and K. Grunewald, Ecosystem service supply and demand — the challenge to balance spatial mismatches. *Int. J. Biodivers. Sci. Ecosyst. Serv. Manag.* **13**(2), 148–161 (2017).

[57] T. Elliot, J. Babí Almenar, S. Niza, V. Proença, and B. Rugani, Pathways to modelling ecosystem services within an urban metabolism framework. *Sustainability* **11**(10), 2766 (2019).

[58] H. H. Khoo, R. M. Eufrasio-Espinosa, L. S. Koh, P. N. Sharratt, and V. Isoni, Sustainability assessment of biorefinery production chains: A combined LCA-supply chain approach. *J. Clean. Prod.* **235**, 1116–1137 (2019).

[59] B. Goldstein, M. Birkved, M. B. Quitzau, and M. Hauschild, Quantification of urban metabolism through coupling with the life cycle assessment framework: Concept development and case study. *Environ. Res. Lett.* **8**(3), 035024 (2013).

[60] A. Petit-Boix, P. Llorach-Massana, D. Sanjuan-Delmás, J. Sierra-Pérez, E. Vinyes, X. Gabarrell *et al.*, Application of life cycle thinking towards sustainable cities: A review. *J. Clean. Prod.* **166**, 939–951 (2017).

[61] J. Albertí, A. Balaguera, C. Brodhag, and P. Fullana-i-Palmer, Towards life cycle sustainability assessment of cities. A review of background knowledge. *Sci. Tot. Environ.* **609**, 1049–1063 (2017).

[62] T. Elliot, J. Babí Almenar, and B. Rugani, Impacts of policy on urban energy metabolism at tackling climate change: The case of Lisbon. *J. Clean. Prod.* **276**, 123510 (2020).

Chapter 13

Advancements in Methods for Life Cycle Assessment of Industrial Symbiosis

Piya KERDLAP[*,‡] **and Jonathan Sze Choong LOW**[†,§]

*National University of Singapore (NUS),
21 Lower Kent Ridge Road, Singapore 119077*

†*Singapore Institute of Manufacturing Technology (SIMTech),
2 Fusionopolis Way, #08-04, Innovis, Singapore 138634*

‡*piyakerdlap@u.nus.edu*

§*sclow@simtech.a-star.edu.sg*

The reallocation of resources among co-located firms involving physical exchanges of materials, energy, water, and by-products is known as "industrial symbiosis", a subfield of "industrial ecology", which has gone through advancements in research and development. This chapter introduces the subject of life cycle assessment (LCA) of industrial symbiosis networks (ISNs). The characteristics of LCAs of ISNs are described and the methodological advancements in this field over the past 10 years are discussed. The models and computations methods used to conduct LCAs of ISNs are summarized along with the different methods that have been used to allocate the benefits and burdens of industrial symbiosis exchanges between companies.

1. Introduction

The field of industrial symbiosis is defined as the engagement of tradition-ally separate industries and entities in a collaborative approach to exchange materials, energy, water, and by-products to create benefits for both the environment and the economy [1]. This concept involves the input–output flow and recycling of materials, resources, and energy in an ecosystem as a potential model for relationships between facilities and firms. The archetypal example is the industrial symbiosis in Kalundborg. Since then, there has been greater interest in further developing such arrangements and establishing eco-industrial parks. Such exchanges in an industrial symbiosis network (ISN) help mitigate waste generation and improve resource efficiency [2–5]. However, other impacts to the environ-ment can occur outside the boundaries of the ISN [6]. This has motivated the application of life cycle assessment (LCA) to holistically quantify the environmental impacts of the interlinked supply chains of an ISN. LCA is used to evaluate the environmental impacts of a product (or service) throughout its entire life cycle [7, 8]. This starts from the extraction of raw materials, manufacturing of materials and products, use of products, reuse, and final disposal at the end of life. The use of LCA to evaluate ISNs has helped quantify the environmental benefits of forming waste-to-resource exchanges between companies and identify potential environ-mental burden shifts. Over the past 10 years, many LCA case studies and methodological studies for ISNs have been conducted. This chapter dis-cusses the methodological advancements in LCA for ISNs.

2. Characteristics of Life Cycle Assessment of Industrial Symbiosis

There are several characteristics that an LCA practitioner usually encoun-ters when modeling and analyzing an ISN. The first characteristic is that there are one or more entities that produce multiple valuable outputs which are referred to as the product(s) and by-product (also referred to as co-product). Since the by-product is valuable, it is converted into a resource that is taken as input into the supply chain of one or more other entities in an ISN. This is also referred to as open-loop recycling. A valu-able waste or by-product can also be converted into a resource and be used within the same supply chain it came from which is referred to as closed-loop recycling.

The second characteristic of LCAs of ISNs is that different resource flows of energy, water, and materials can have multiple origins (sources) and destinations (sinks). In ISNs, two or more resource flows may be physically identical, but need to be differentiated because they have different sources and sinks. An example is when two waste flows are the same physical substance, but one waste flow is consumed by a recycling process and the other is disposed at a landfill or incinerator.

Similar to conducting LCAs of single-product systems, LCAs of ISNs have foreground and background systems. The foreground system typically includes processes that take place within the companies in the ISN and waste-to-resource exchanges between companies. The background system usually includes upstream processes such as extraction and production of raw materials required by the individual companies as well as treatment of any wastes produced by the ISN.

3. Timeline of Methodological Advancements in Life Cycle Assessment of Industrial Symbiosis Networks

Over the past decade, there has been a wide range of advancements in LCA methods used to evaluate the environmental performance of ISNs. These methodological advancements are illustrated in a timeline in Figure 1. The studies shown in the timeline have made contributions in various areas which comprise guidance in designing LCA studies to answer different types of questions about an ISN, models and computational methods, and methods for allocating the environmental benefits and burdens of waste-to-resource exchanges between different companies in an ISN.

3.1. *Models and computational methods*

A total of eight unique methodologies for modeling and analyzing the life cycle environmental performance of ISNs have been developed between 2010 and 2020 [9, 10]. The main differentiation between each of the methodologies is how the LCA foreground and background systems of an ISN are represented. The methodologies can generally be grouped into two categories of process-based approach and non-process-based approach. Within the category of process-based approach is the

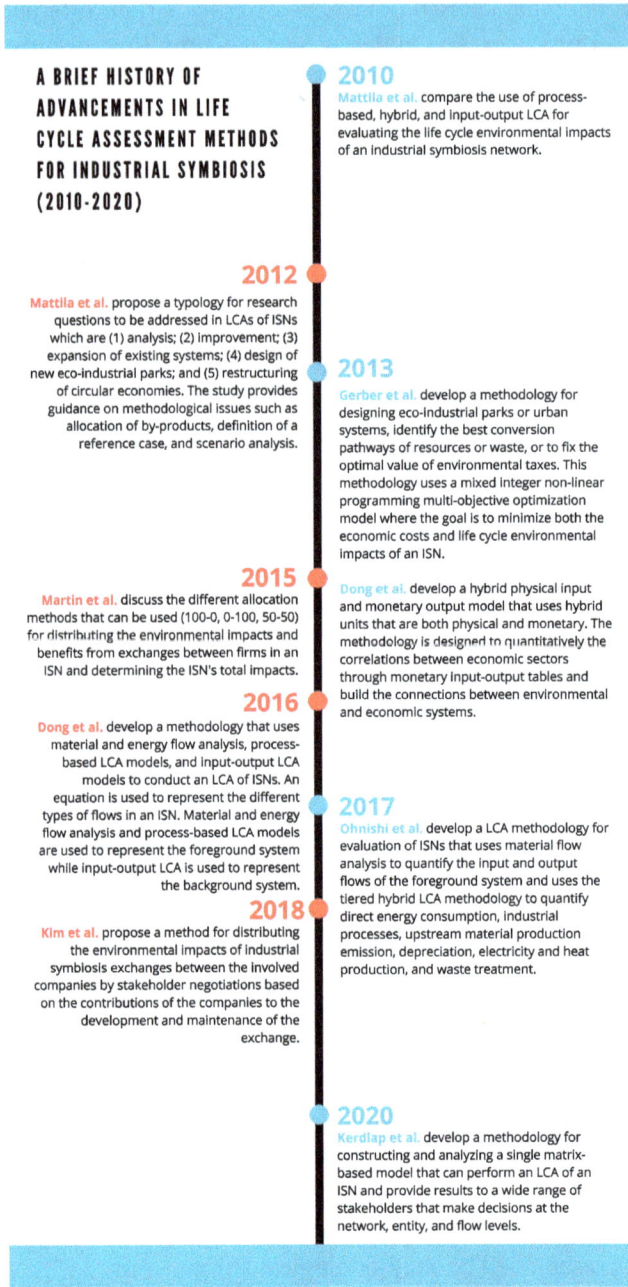

A BRIEF HISTORY OF ADVANCEMENTS IN LIFE CYCLE ASSESSMENT METHODS FOR INDUSTRIAL SYMBIOSIS (2010-2020)

2010
Mattila et al. compare the use of process-based, hybrid, and input-output LCA for evaluating the life cycle environmental impacts of an industrial symbiosis network.

2012
Mattila et al. propose a typology for research questions to be addressed in LCAs of ISNs which are (1) analysis; (2) improvement; (3) expansion of existing systems; (4) design of new eco-industrial parks; and (5) restructuring of circular economies. The study provides guidance on methodological issues such as allocation of by-products, definition of a reference case, and scenario analysis.

2013
Gerber et al. develop a methodology for designing eco-industrial parks or urban systems, identify the best conversion pathways of resources or waste, or to fix the optimal value of environmental taxes. This methodology uses a mixed integer non-linear programming multi-objective optimization model where the goal is to minimize both the economic costs and life cycle environmental impacts of an ISN.

2015
Martin et al. discuss the different allocation methods that can be used (100-0, 0-100, 50-50) for distributing the environmental impacts and benefits from exchanges between firms in an ISN and determining the ISN's total impacts.

Dong et al. develop a hybrid physical input and monetary output model that uses hybrid units that are both physical and monetary. The methodology is designed to quantitatively the correlations between economic sectors through monetary input-output tables and build the connections between environmental and economic systems.

2016
Dong et al. develop a methodology that uses material and energy flow analysis, process-based LCA models, and input-output LCA models to conduct an LCA of ISNs. An equation is used to represent the different types of flows in an ISN. Material and energy flow analysis and process-based LCA models are used to represent the foreground system while input-output LCA is used to represent the background system.

2017
Ohnishi et al. develop a LCA methodology for evaluation of ISNs that uses material flow analysis to quantify the input and output flows of the foreground system and uses the tiered hybrid LCA methodology to quantify direct energy consumption, industrial processes, upstream material production emission, depreciation, electricity and heat production, and waste treatment.

2018
Kim et al. propose a method for distributing the environmental impacts of industrial symbiosis exchanges between the involved companies by stakeholder negotiations based on the contributions of the companies to the development and maintenance of the exchange.

2020
Kerdlap et al. develop a methodology for constructing and analyzing a single matrix-based model that can perform an LCA of an ISN and provide results to a wide range of stakeholders that make decisions at the network, entity, and flow levels.

Figure 1.   Timeline of methodological advancements in LCA of ISNs.

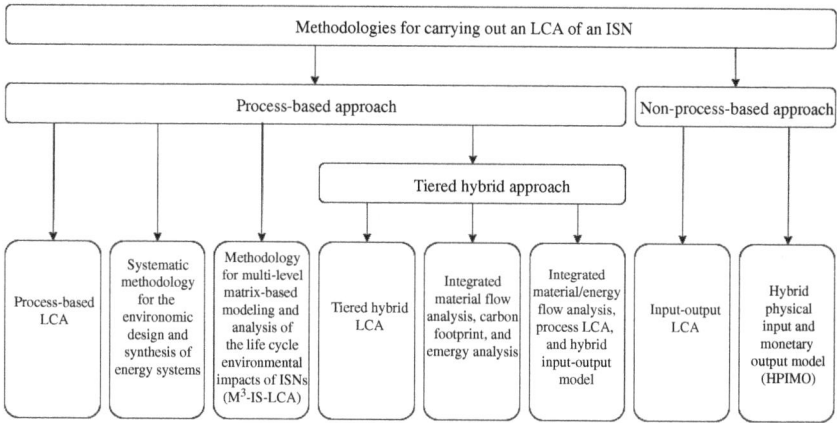

Figure 2. Categorization of methodologies for LCAs of ISNs.

sub-category of the tiered hybrid approach. This categorization is illustrated in Figure 2.

In the process-based approach, the life cycle inventory is constructed by identifying all the different processes that take place in the supply chain. Process-specific data is used to represent either some stages of the life cycle or all stages of the life cycle. The non-process-based approach uses environmentally extended input–output analysis. In this type of analysis, aggregated economic sector input–output data is used to quantify the inputs and outputs of an activity in the LCA. The LCA methodologies in the category of tiered hybrid approach use process-specific data or material flow analysis in the foreground system and use aggregated economic input–output data in the background system.

The process-based LCA methodology was used in 77% of LCA studies of ISNs conducted between 2010 and 2019, while the remaining studies used tiered hybrid LCA and other methodologies [9]. One of the reasons studies have used LCA methodologies that take a tiered hybrid approach was that it was difficult to access process-specific inventory data in the study's respective countries to represent upstream material supply and transport in the background system [11–13]. Therefore, the use of a pure process-based LCA model was not able to help complete the analysis.

Each of the methodologies is able to conduct multi-level analysis to provide LCA results for a wide range of ISN stakeholders that make decisions at different levels of the ISN. This multi-level LCA is important because stakeholders at different levels will have their own unique set of goals and objectives. For instance, when policymakers are evaluating incentives for fostering waste-to-resource exchanges among companies in an ISN, aggregated information at the network level of how the overall ISN could perform if the incentives were implemented would be needed. For companies, insights at the entity level on whether participating in an ISN adds value to their business would be more meaningful. An even higher resolution at the product or resource flow level is required when product designers are deciding on the materials to use in their products.

Depending on the structure of the LCA model and the number of models used to represent the foreground and background systems, different methodologies need either single or multiple computations to produce the LCA results at the network, entity, and resource flow levels. Both the matrix-based model used in process-based LCAs and the input–output LCA methodology can analyze a single or a bundle of products in an ISN through constructing a single matrix. For the case of the tiered hybrid LCA methodology, two or more computations are needed to carry out multi-level LCAs because different LCA models and computational methods are used in the foreground and background system. The first computation for the foreground system would be done with either the matrix-based model or sequential equations. The second computation for the background system would be done through an input–output LCA model where aggregated industry sector data are used instead of process-based inventory data. There are other hybrid LCA methodologies that can carry out multi-level analysis in a single computation such as the integrated hybrid LCA and the matrix augmentation methodology (also often referred to as input–output-based hybrid). However, these two hybrid LCA methodologies have not yet been used to carry out LCAs of ISNs to date [9]. To address this specific modeling challenge, Kerdlap *et al.* [10] developed a methodology for constructing a single matrix-based model that can provide LCA results about an ISN at multiple levels. The methodology outlines a formalism for constructing a matrix-based LCA model that is consistently a square dimension and invertible. Through manipulation of the demand vector, the model can be analyzed to provide LCA results about the ISN at the network, entity, and flow levels.

3.2. *Allocation methods*

Waste-to-resource exchanges are a fundamental characteristic of ISNs which result in avoiding the use of raw materials by a company that uses the waste or by-product. Computing the change in environmental impacts of an ISN from the perspective of the whole network before and after waste-to-resource exchanges is fairly straightforward. The challenge arises when the network-level impacts need to be distributed among the different companies involved. This is because, in a waste-to-resource exchange, there is usually a process that incurs environmental impacts for converting the waste into a valuable resource. The valuable resource is then consumed by another company and displaces a certain amount of resource that originally came from a virgin source. The question that arises is how should the environmental impacts of converting the waste into a resource and the avoided environmental impacts of replacing a raw material be allocated among the companies involved in the waste-to-resource exchange? To date, there is no correct method to solve this problem, but the allocation choice should be consistent with the goal and scope of the study and its methodological choices [14]. LCA case studies by Martin *et al.* [15, 16], Kim *et al.* [17] and Vigano *et al.* [18] have shown that the choice of allocation method affects the LCA results for individual companies involved in an ISN. The three methods that have been used for allocation are 100–0 (also referred to as cutoff allocation), 0–100 (also referred to as substitution), and 50/50 which are illustrated in Figure 3. Each method can be summarized as follows:

- *100–0 method*: All the credits for avoided resource requirements are given to the company that produced the by-product that could be used as a resource for the receiving company.
- *0–100 method*: The company that receives the by-product gets the credit for avoided resource use.
- *50/50 method*: The credits are evenly split between both companies involved in the waste-to-resource exchange.

Kim *et al.* [17] proposed a method where the credits are negotiated between the companies based on contributions of the companies. This negotiation would be overseen by a third party such as a greenhouse gas accreditation agency.

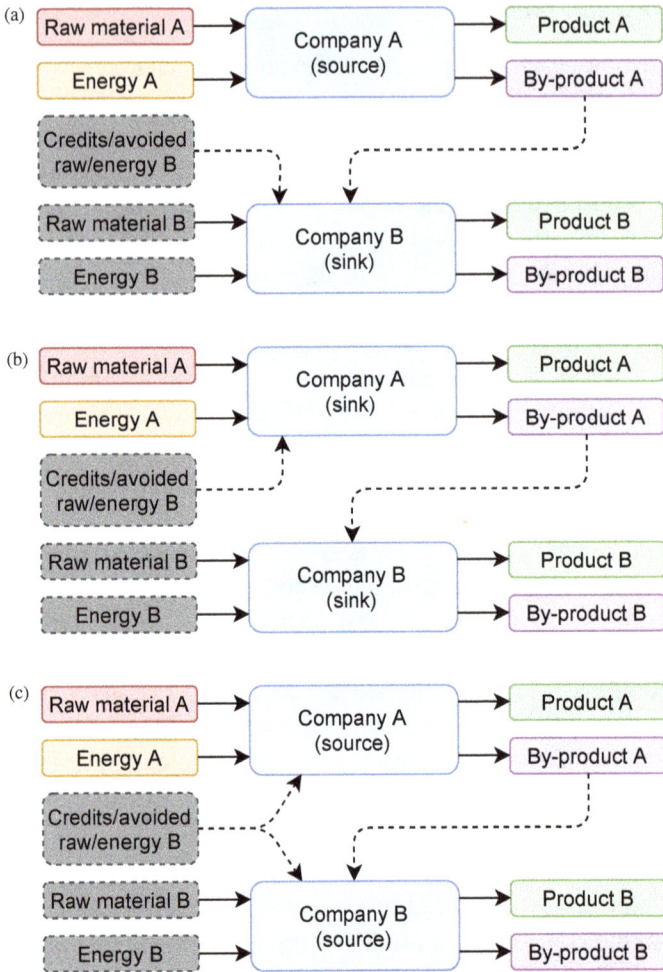

Figure 3.   Allocation methods for exchanges between company A and company B through (a) 100–0, (b) 0–100, and (c) 50/50 methods.

4. Conclusion

LCA is an essential tool for holistically evaluating the environmental performance of ISNs. A holistic approach helps in understanding whether a proposed technical solution applied locally achieves the desired environmental benefits in a broader context. In the case of ISNs, LCAs help quantify the environmental benefits and burdens of waste-to-resource exchanges and determine how they affect the entire network and the participating companies. This helps to mitigate burden shifts that can lead to a circular economy rebound effect [19].

Many methodological developments have taken place from 2010 to 2020 in the areas of models and computational methods and allocation methods. Examples of intensive work to expand LCA methodologies to better address the environmental impacts associated with industrial symbiosis have been ongoing and have been steadily increasing [20–22]. As one example, Mattila *et al.* [23] compared process, hybrid, and input–output LCA approaches in quantifying the overall environmental impacts of a forest industrial symbiosis, situated in Kymenlaakso, Finland. In another noteworthy example, Aissani *et al.* [24] highlighted methods and parameters used to define and design a reference scenario to be compared with an industrial symbiosis scenario using LCA methodology. The authors examined a total of 26 peer-reviewed papers that used LCA in the field of ISNs.

A majority of LCA studies of ISNs have used the process-based LCA model, while tiered hybrid LCA methodologies were the second most used. All the LCA methodologies are capable of carrying out multi-level analysis. However, each methodology differs in the number of computations required to construct and analyze the LCA model to provide results at different levels. Different allocation methods exist for distributing the benefits of waste-to-resource exchanges among the participating companies in an ISN. The allocation method chosen should be in line with the goal and scope of the LCA study and be agreed upon between the stakeholders involved.

Future areas of research lie in the development of models that unify LCA and the life cycle cost analysis of ISNs. Such work has been completed for single-product systems, but not for the case of multi-product systems that occur in ISNs. Research in this area could support industrial symbiosis facilitation tools in carrying out simultaneous life cycle environmental and economic evaluations of the potential symbiosis connections identified.

References

[1] M. R. Chertow, Industrial symbiosis: Literature and taxonomy. *Annu. Rev. Energy Environ.* **25**, 313–337 (2000).

[2] J. S. C. Low, T. B. Tjandra, F. Yunus, S. Y. Chung, D. Z. L. Tan, B. Raabe *et al.*, A Collaboration platform for enabling industrial symbiosis: Application of the database engine for waste-to-resource matching. *Procedia CIRP* **69**, 849–854 (2018).

[3] Z. Yeo, J. S. C. Low, D. Z. L. Tan, S. Y. Chung, T. B. Tjandra, and J. Ignatius, A collaboration platform for enabling industrial symbiosis: Towards creating a self-learning waste-to-resource database for recommending industrial symbiosis transactions using text analytics. *Procedia CIRP* **80**, 643–648 (2019).

[4] Z. Yeo, D. Masi, Y. T. Ng, P. S. Tan, and S. Barnes, Tools for promoting industrial symbiosis: A systematic review. *J. Ind. Ecol.* **23**, 1087–1108 (2019).

[5] P. Kerdlap, J. S. C., Low, and S. Ramakrishna, Zero waste manufacturing: A framework and review of technology, research, and implementation barriers for enabling a circular economy transition in Singapore. *Res. Conserv. Rec.* **151**, 104438 (2019).

[6] P. Kerdlap, J. S. C. Low, R. Steidle, D. Z. L. Tan, C. Herrmann, and S. Ramakrishna, Collaboration platform for enabling industrial symbiosis: Application of the industrial-symbiosis life cycle analysis engine. *Procedia CIRP* **80**, 655–660 (2019).

[7] ISO 14040, *Environmental Management: Life Cycle Assessment — Principles and Framework* (International Organization for Standardization, Geneva, 2006).

[8] ISO 14044, *Environmental Management: Life Cycle Assessment — Requirements and Guidelines* (International Organization for Standardization, Geneva, 2006).

[9] P. Kerdlap, S. C., Low, and S. Ramakrishna, Life cycle environmental and economic assessment of industrial symbiosis networks: A review of the past decade of models and computational methods through a multi-level analysis lens. *Int. J. Life Cycle Assess.* **25**, 1660–1679 (2020).

[10] P. Kerdlap, J. S. C. Low, D. Z. L. Tan, Z. Yeo, and S. Ramakrishna, M3-IS-LCA: A methodology for multi-level life cycle environmental performance evaluation of industrial symbiosis networks. *Res. Conserv. Rec.* **161**, 104963 (2020).

[11] H. Dong, Y. Geng, F. Xi, and T. Fujita, Carbon footprint evaluation at industrial park level: A hybrid life cycle assessment approach. *Energ. Policy* **57**, 298–307 (2013).

[12] L. Dong, T. Fujita, M. Dai, Y. Geng, J. Ren, M. Fujii *et al.*, Towards preventive eco-industrial development: An industrial and urban symbiosis case in one typical industrial city in China. *J. Clean. Prod.* **114**, 387–400 (2016).

[13] L. Dong, H. Liang, L. Zhang, Z. Liu, Z. Gao, and M. Hu, Highlighting regional eco-industrial development: Life cycle benefits of an urban industrial symbiosis and implications in China. *Ecol. Model.* **361**, 164–176 (2017).

[14] J. B. Guinée, R. Heijungs, and G. Huppes, Economic allocation: Examples and derived decision tree. *Int. J. Life Cycle Assess.* **9**, 23–33 (2004).

[15] M. Martin, Quantifying the environmental performance of an industrial symbiosis network of biofuel producers. *J. Clean. Prod.* **102**, 202–212 (2015).

[16] M. Martin, N. Svensson, and M. Eklund, Who gets the benefits? An approach for assessing the environmental performance of industrial symbiosis. *J. Clean. Prod.* **98**, 263–271 (2015).

[17] H.-W. Kim, S. Ohnishi, M. Fujii, T. Fujita, and H.-S. Park, Evaluation and allocation of greenhouse gas reductions in industrial symbiosis. *J. Ind. Ecol.* **22**, 275–287 (2018).

[18] E. Viganò, C. Bondi, S. Cornago, A. Caretta, L. Bua, L Carnelli *et al.*, The LCA modelling of chemical companies in the industrial symbiosis perspective: Allocation approaches and regulatory framework. In: M. Simone and C. Brondi (Eds.), *Life Cycle Assessment in the Chemical Product Chain* (Springer, Switzerland, 2020), pp. 75–98.

[19] T. Zink and R. Geyer, Circular economy rebound. *J. Ind. Ecol.* **21**, 593–602 (2017).

[20] K. Soratana and A. E. Landis, Evaluating industrial symbiosis and algae cultivation from a life cycle perspective. *Bioresour. Technol.* **102**(13), 6892–6901 (2011).

[21] L. Sokka, S. Lehtoranta, A. Nissinen, and M. Melanen, Analyzing the environmental benefits of industrial symbiosis life cycle assessment applied to a Finnish forest industry complex. *J. Ind. Ecol.* **15**, 137–155 (2011).

[22] D. F. Vivanco and E. van der Voet, The rebound effect through industrial ecology's eyes: A review of LCA-based studies. *Int. J. Life Cycle Assess.* **19**, 1933–1947 (2014).

[23] T. J Mattila, S. Pakarinen, and L. Sokka, Quantifying the total environmental impacts of an industrial symbiosis — A comparison of process-, hybrid and input-output life cycle assessment. *Environ. Sci. Technol.* **44**(11), 4309–4314 (2010).

[24] L. Aissani, A. Lacassagne, J.-B. Bahers, and S. Le Féon, Life cycle assessment of industrial symbiosis: A critical review of relevant reference scenarios. *J. Ind. Ecol.* **23**, 972–985 (2019).

Chapter 14

An Integrated Techno-Sustainability Assessment: Methodological Guidelines

Van Schoubroeck SOPHIE[*,¶], Thomassen GWENNY[*,†,||], Compernolle TINE[*,‡,**], Van Dael MIET[*,§,††] and Van Passel STEVEN[*,‡‡]

[*]*University of Antwerp, Prinsstraat 13, 2000 Antwerp, Belgium*

[†]*Ghent University, Research Group Sustainable Systems Engineering, Coupure Links 653, 9000 Ghent, Belgium*

[‡]*Royal Belgian Institute of Natural Sciences, Vautierstraat 29, 1000 Brussels, Belgium*

[§]*VITO, Unit Separation and Conversion Technologies, Boeretang 200, 2400 Mol, Belgium*

[¶]*sophie.vanschoubroeck@uantwerpen.be*

[||]*gwenny.thomassen@uantwerpen.be*

[**]*tine.compernolle@uantwerpen.be*

[††]*miet.vandael@vito.be*

[‡‡]*steven.vanpassel@uantwerpen.be*

This chapter introduces the integrated techno-sustainability assessment (TSA) framework, combining environmental, economic, and social impact assessments for emerging technologies and products. Following

245

the life cycle sustainability approach (LCSA), TSA tackles the current challenges regarding indicator selection, quantification, and integration. Methodological guidelines are provided to cope with assessments of technologies and products in the early development stages, which contain many uncertainties and data gaps. TSA aims to guide decision makers throughout the different technology-readiness levels (TRLs), by providing information to make better-informed choices toward sustainability and improve technological parameters while changes can still be made.

1. Introduction

The concept of "sustainable development", which can be interpreted as process-based pathways or new technologies for new product design to achieve the sustainability goals, was introduced first in 1980 by the International Union for the Conservation of Nature and Natural Resources [1]. The popularity of the concept increased after the publication of Our Common Future in 1987. In 2002, at the World Summit on Sustainable Development, the United Nations (UN) first introduced the three dimensions concept for sustainable development, which embraces economic development, social development, and environmental protection. Sustainability assessment approaches that evaluate all three dimensions, preferably in an integrated and connecting way, are needed. To move beyond an individual, disconnected economic, environmental, or social assessment, life cycle sustainability assessment (LCSA) was introduced. Kloepffer [2] described the LCSA framework as a combination of life cycle costing (LCC) to determine the economic impact, life cycle assessment (LCA) for the environmental impact, and social life cycle assessment (social LCA) for the social impact (Equation (1)). The UNEP formalized the LCSA concept in 2011 in a report called "Towards a Life Cycle Sustainability Assessment" [3]. After the publication of this methodological report, the number of peer-reviewed articles that further develop and apply LCSA has increased (see Figure 1).

$$LCSA = LCC + LCA + social\ LCA \tag{1}$$

LCSA has received increasing attention over the past years, while at the same time challenges still exist to define the content of the three

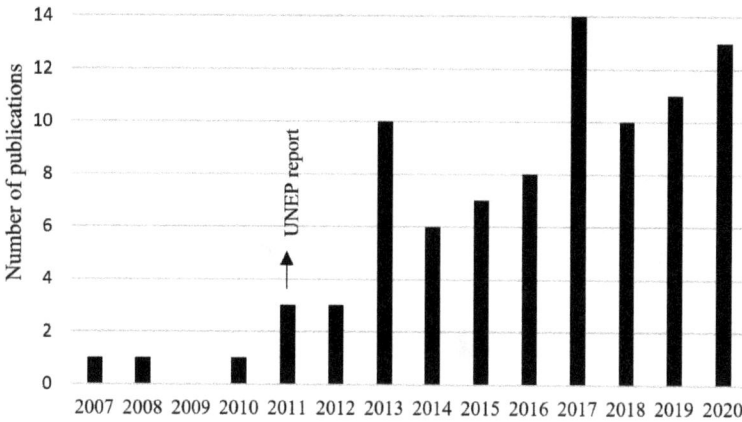

Figure 1. Number of publications up to and including 2020. A search in the Science Citation Index Expanded for peer-reviewed articles, containing the phrases "life cycle sustainability assessment" (82) or "life cycle sustainability analysis" (5) or "life cycle sustainability evaluation" (1). In total, 88 publications were found.

sustainability dimensions, integrate them, and effectively communicate LCSA application results [4, 5]. Multiple reasons exist for this, such as the presence of sector- or product-specific characteristics which cannot be generalized or the value-loaded nature of what the different sustainability domains entail. Second, to obtain a fully integrated assessment, harmonization of environmental, economic, and social assessment techniques is needed. In other words, assessment characteristics such as system boundaries and functional units should be aligned in LCSA applications. Hoogmartens *et al.* [6] clarified the key relations between the different assessment techniques and showed how dissimilar approaches interact. Third, the inclusion of multiple dimensions automatically implies a multitude of indicators being involved, which are often expressed in different units. When assessing all the indicators in one assessment, the risk exists to support the weak sustainability view that allows the possibility for trade-offs. This raises some ethical questions concerning the allowance of specific trade-offs, such as the trade-off between profit (i.e. an economic gain) and a fair income (i.e. a social concern). Also, this full compensation could be in direct conflict with the planetary boundaries and social thresholds on a macro level, which are described in models, such as the

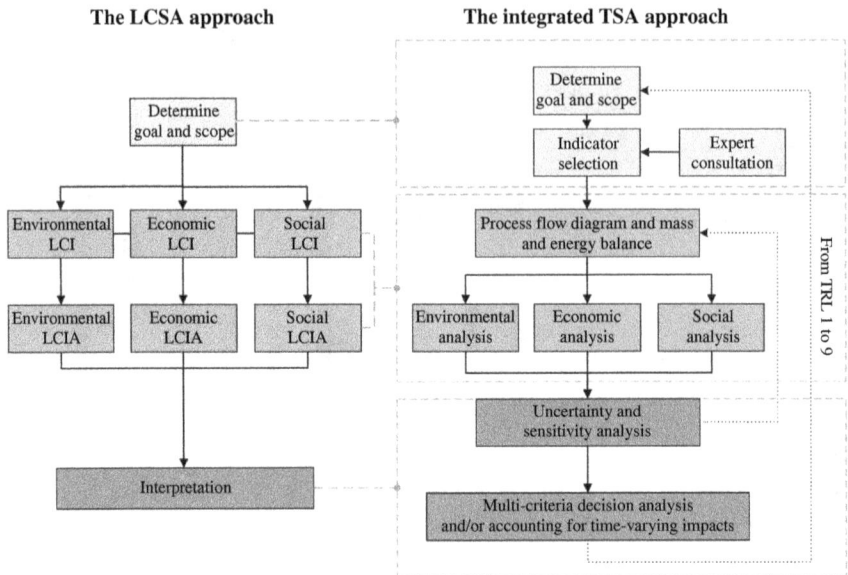

Figure 2.   A TSA for different TRLs, following an LCSA approach.

Doughnut economy [7]. Lastly, the current LCSA approach is focused on ex-post assessments, while the *ex-ante* evaluation of new technologies and products could identify important hotspots to prevent large environmental costs, economic losses, and social irregularities already at low technology-readiness levels (TRLs) when there is still greater flexibility toward improvements [8].

Given the current challenges regarding LCSA and its focus on an *ex-post* approach, the integrated techno-sustainability assessment (TSA) framework was introduced to assess technologies and products already from early development stages and provide a practical approach to integrate all three sustainability dimensions (see Figure 2) [9]. In the following sections, the integrated TSA will be introduced, and methodological guidelines are provided for a comprehensive indicator selection, quantification, and integration for sustainable decision-making, applicable to technologies in the early development stages, such as emerging CO_2 utilization technologies (Chapter 3). This chapter aims to provide a methodological framework that can be implemented to obtain a fully integrated sustainability assessment, which covers environmental, economic, and social impacts.

2. Assessment's Goal and Scope

The integrated TSA methodology starts by defining the goal and scope of the assessment, which is also the first step in the LCSA approach (see Figure 2). Before sustainability indicators are selected and impact calculations are initiated, the evaluator should define "what" needs to be assessed. Three major parts of this are the delineation of the system boundaries, the choice regarding the functional unit, and the temporal and geographical scale of the assessment. The system boundaries identify the unit processes that are evaluated in the concerning study. The choice of these boundaries depends on the goal of the assessment and in most cases also on the availability of data. The functional unit provides a reference to which the inputs and outputs can be related. When comparing multiple scenarios within a sustainability evaluation, the same functional unit is needed to make valid comparisons. The delineation and definition of the system boundaries and the functional units are common practice in LCA (Chapter 1). The definition of the temporal and geographical scale is an important consideration to gather the appropriate data in the life cycle inventory step for LCA. However, this is also an essential part of the scope definition for LCC and social LCA as the economic and social impacts are often even more dependent on temporal and geographical considerations compared to environmental impacts. When including an environmental, economic, and social evaluation in the goal of the assessment, there should be consistency for the system boundaries and scale for the whole integrated assessment [4]. In general, one should avoid approaching a technology or activity without considering all its upstream and downstream value chain activities. A life cycle view should be supported during the sustainability assessment.

3. Indicator Selection

An indicator-based method is considered as being flexible to quantify a range of economic, social, and environmental impacts, complying with the three sustainability dimensions [10]. In the second step of the TSA framework, a range of indicators is identified to measure the sustainability performance [9, 11].

The first challenge to follow an indicator-based approach pertains to the translation of the three sustainability dimensions into measurable indicators that are relevant to the assessed technology or product. These indicators are used to assess and evaluate performances, to provide trends on improvements and highlight sustainability "hotspots" along the product life cycle. They provide information to decision makers to formulate strategies and communicate potential achievements to stakeholders [12]. However, an extensive indicator selection, referring to all relevant sustainability impacts, is often lacking when performing a sustainability analysis. "Popular" indicators, such as Global Warming Potential (GWP) as an environmental indicator or the cost of raw materials as an economic indicator, are naturally considered by many assessments without consulting stakeholders for the inclusion of other relevant impacts [13]. The nature and content of a sustainability assessment are case-specific, and indicator sets should be defined accordingly. The selection of sustainability indicators determines the assessment results. For that reason, an inconsistent selection of indicators could lead to incomparable results within or between projects [14].

When developing sets of indicators, a thorough understanding of the market and policy environment is necessary. Two approaches for indicator selection exist, that is, a literature review or a participatory (stakeholder) approach [15]. Design validation of the indicators can be increased by using expert judgments for their selection [16]. Expert involvement could aid in the selection of indicators combined with a prioritization exercise to determine the relative importance of each [17]. There are multiple ways to structure this stakeholder participation, such as brainstorm sessions, focus groups, or questionnaires. The latter could be executed through online surveys with experts and carried out employing a Delphi method, "a method used to obtain the most reliable consensus or opinion of a group of experts by a series of intensive questionnaires interspersed with controlled feedback" [18]. As the definition explains, this Delphi method is specifically designed to reach consensus and is generally known to have a flexible design that can be adapted to the specific study needs [19].

Next to a selection of indicators, the participatory approach can also be used to develop a ranking of the selected indicators. A prioritization of indicators is useful when resources, such as data and time, are limited [20]. This often occurs when technologies are assessed in the early development stages or for projects in small- and medium-sized enterprises (SMEs). Besides, a ranking facilitates decision-making by providing information on possible weights, indicating the relative degree of importance of the indicators. Data to develop these rankings are retrieved by applying research techniques, such as choice modeling (e.g. best-worst scaling) or rating (e.g. using the Likert scale). The question on which technique is preferred will depend on the needs of the researcher and case-specific characteristics such as the number of indicators to be ranked and the geographical location of the experts involved (e.g. to avoid scale use bias). Another option would be to assess the level of agreement of experts with existing ranked indicator sets. Rank correlation coefficients, such as Kendall's tau and Spearman's rho can be used to evaluate and compare existing rankings with new ones [21, 22]. More information on the construction of both rank correlation coefficients can be found in de Keyser and Springael [23]. Bengtsson and Steen described a variety of existing weighting methods for LCA, which can contribute to sustainable decision-making and the relevance and acceptability of the LCA results [24].

4. Sustainability Analysis

4.1. *Process flows and mass and energy balance*

To be able to identify the improvement potential of an emerging technology, a detailed technological assessment, extending further than the traditional listing in a life cycle inventory as is common practice in LCSA, is required. First, a corresponding process flow diagram is drawn. Relevant technological data are collected concerning the processes and products one wants to assess. With that information, the mass and energy flows are calculated (Chapter 1). The nature of the data and level of detail are highly dependent on the TRL of the considered technology, ranging from an initial idea situated at TRL 1 or 2 to a proven and mature technology at TRL 9 (see Figure 3) [25].

The first quantitative streamlined assessment is possible around TRL 3 or 4 [26]. When assessing emerging technologies, from TRL 3 to

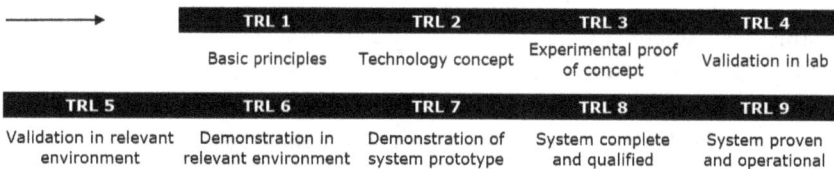

TRL 1	TRL 2	TRL 3	TRL 4
Basic principles	Technology concept	Experimental proof of concept	Validation in lab

TRL 5	TRL 6	TRL 7	TRL 8	TRL 9
Validation in relevant environment	Demonstration in relevant environment	Demonstration of system prototype	System complete and qualified	System proven and operational

Figure 3. An overview of TRLs, defined by the European Commission.

TRL 6, information sources are often limited to lab-scale results and primary pilot data, computer models, and secondary data from the literature. These types of data entail many uncertainties, and many process operations and parameters change while scaling up technologies. Therefore, scale-up calculations might be required, such as the ones introduced by Piccino *et al.* [27]. A more extensive overview of how to deal with this prospective perspective can be found in various prospective LCA review studies [26, 28–30].

4.2. *Indicator measurements*

After the scope of the assessed system is defined and the mass and energy balances are calculated, the identified sustainability indicators should be measured and linked to the technical parameters as much as possible. Following the three dimensions approach of sustainability, one should measure environmental, economic, and social aspects.

4.2.1. *Environmental assessment*

The LCA method is a globally recognized method for environmental analysis over the entire life cycle of a product and follows the ISO guidelines 14040–14044 ISO reference. Of the three sustainability dimensions, the environmental impact assessment using an LCA has gained the most attention and has already been presented by other authors [6, 10, 16] and therefore will not be discussed in detail in this section. LCA methodologies and developments for various applications can also be found in various other chapters of this book.

4.2.2. *Economic assessment*

Within the economic dimension, two popular methods can be distinguished: the techno-economic assessment (TEA) and the life cycle costing assessment (LCC). A TEA focuses on the perspective of the technology developer and aims to assess the economic feasibility and identifies technological and non-technological barriers by a risk analysis to define clear technological development targets [31]. The most efficient pathways for technology development are mapped by directly linking (i.e. integrating) technological and economic parameters. As a TEA aims to provide valuable information to a process developer, it follows by nature an ex-ante approach. On the other hand, LCC is focused on the cost distributions of a product, considering all phases of its life cycle (described in Chapter 9). Similar to LCA, LCC was constructed for ex-post analysis. Despite the different perspectives of both assessments, the underlying methodology for the economic calculations is the same for both TEA and LCC and therefore the methodologies can, but should not, be different in the choice of the goal and scope [32].

Research guidelines for specific TEA applications (e.g. for CO_2 utilization) do exist but are also criticized for not considering maturity levels and the dynamics of emerging technologies [33–35]. A general guideline for executing TEA studies was provided by Van Dael *et al.* [36]. Examples of recent TEA studies can be found in Larrain *et al.* [37] and Tschulkow *et al.* [38] concerning the mechanical recycling of plastic packaging waste and biorefinery processes. To facilitate the ex-ante economic and environmental assessment of emerging technologies, the TEA method has been integrated with a prospective LCA by Thomassen *et al.* [39] and can be referred to as an environmental techno-economic assessment (ETEA).

4.2.3. *Social assessment*

Social impact assessment (SIA) is defined by the International Association for Impact Assessment (IAIA) as "the process of analyzing, monitoring, and managing the intended and unintended social consequences, both positive and negative, of planned interventions (policies, programs, plans, and projects) and any social change processes invoked by these interventions" [40]. The addition of the social dimension completes the three-dimensions sustainability framework and is most often approached by the social life cycle assessment (social LCA) method. UNEP and SETAC have set up a Life Cycle Initiative supporting social LCA by developing methodological sheets and proposing social indicators for different stakeholder groups (including workers, local communities, etc.) [41]. The recurring problem herein is the lack of research and quantification methods for the assessment of these indicators [42]. One of the challenges highlighted for defining social LCA indicators lies within the fact that location information varies significantly in cultural, demographic, and economic disparities between countries [43]. The risk exists that qualitative features in social LCA remain "sidelined" and will only serve as descriptive context when assessing the full sustainability of technologies and products [44]. Social indicators that are usually qualitative can be converted to semi-quantitative ones by introducing a scoring strategy [45]. Stakeholder surveys can be conducted to gather the necessary information, for example on indicators such as community involvement, discrimination, and gender equality [46]. However, an ex-ante social assessment will often be limited to only a few indicators or based on sector or country averages [46]. While moving toward a full-scale technology, data on the behavior of existing companies can be used to evaluate, monitor, and improve social impacts in social LCA [47].

Previous research showed that the acceptance of technologies and products is important to assess, definitely when evaluating emerging technologies in an ex-ante assessment [20]. While social impacts measure the consequences of actions, social acceptance measures if an innovation will be accepted (or rejected) by key actors [48]. Low social acceptance often results in delays of both project approval and associated network infrastructure. Such delays do not only increase costs but also put at risk the targets to which policymakers have committed themselves. The social outcomes of any technology adoption may result in an unequal distribution of costs, risks, and benefits associated with the new technologies. Therefore, it is important to address concerns about the siting of new systems (e.g. renewable energy technologies), concerns for procedural fairness (if and how communities are engaged fairly in decision-making), and the need to engage different actors, irrespective of existing capabilities and external factors like information, communication, and knowledge [49]. Such multi-layer assessments could not only improve decision-making processes but also provide insights to embed values of environmental justice in the technology designs. Oosterlaken [50], for instance, applied a Value Sensitive Design (VSD) approach to wind power production with the aim to better reflect key stakeholder values concerning wind turbines and parks. However, it is rather difficult to express the expected effects of technology developments on environmental justice in quantitative terms and to integrate these aspects within TSA. The effects are expected to be diverse because various stakeholder groups are involved, which makes them challenging to compare or be made commensurable by expressing them in the same unit of measurement [51].

Therefore, the appraisal of the acceptance of technology developments should be considered using a wide range of evaluative perspectives, participants, and assessment criteria. A constructive sustainability assessment (CSA) was proposed by Matthews *et al.* [52] to blend life cycle thinking with principles of responsible research and innovation. A combination of research methods can be used — stakeholder value mapping [53], Q methodology [54], Delphi [55], discrete choice experiments [56], and MCDA [51] — to specify the relations between different stakeholders involved, to explore their rationales and discourses on issues of environmental justice, and to compare and weigh up different types and qualities of information on the social dimension of technology development.

5. Uncertainty and Sensitivity Analysis

When all sustainability indicators are measured, one can start interpreting the results. Typically, uncertainty and sensitivity analyses are performed to deal with the uncertainty surrounding the data and to identify the most influential parameters in the study. When assessing emerging technologies in the early development stages, uncertainties in technology performance and costs are often substantial. These uncertainties should be analyzed and reported to provide a robust understanding to decision makers [57].

Uncertainty analysis can map the uncertainties related to the input data and the corresponding effect on the sustainability indicator. A sensitivity analysis is an iterative process that assesses the specific impact of the uncertain input parameters on the model results. When the real uncertainty distribution for all data is used, a global sensitivity analysis is performed. But the challenge remains to identify the ranges and distribution functions for the input parameters, definitely in the early TRL stages [26]. For this reason, some rules of thumb have been introduced to define the level of uncertainty, such as the approach of Brun *et al.* (2002) or the pedigree matrix [58, 59]. Van der Spek [57] performed a critical review of different uncertainty analysis methods and additional guidelines for their use.

An additional and highly relevant aspect is the inclusion of learning effects in prospective technology assessments. Thomassen *et al.* [60]

provided five recommendations that lead to an improved perspective on the environmental impact and cost structure of new technologies and a better comparison base between emerging and conventional technologies.

6. Decision-Making Under Uncertainty

6.1. *Multi-criteria decision-making*

In the previous sections of this chapter, guidelines were provided to select and quantify sustainability indicators from different sustainability domains. Next, the TSA results can be further integrated and used for strategic decision-making. When this integration step is added, the TSA framework is further referred to as an "integrated" techno-sustainability assessment. Multi-criteria decision analysis (MCDA) is a popular method that can deal with the following: (1) multiple attributes from different dimensions (i.e. the list of indicators) and the variety of units in which the values of the attributes are expressed and (2) the subjectivity related to the importance of these attributes. Whenever the attributes are selected and defined, a general decision matrix is developed, summarizing the necessary information for further decision-making. The decision matrix D includes the evaluation of scenario i based on an indicator j, where $i = 1,2,\ldots,$ m, and $j = 1,2,\ldots,$ n (see Figure 4). Weights are attached to the indicators, which reflect the (relative) importance of the indicators. An MCDA can be executed using different sub-methods to define weights and perform aggregation, such as (fuzzy) AHP, TOPSIS, PROMETHEE, or ELECTRE. The appropriateness and applicability of these methods can change according to the goal of the assessment, the nature of the criteria

$$D = \begin{Bmatrix} I_1(SC_1) & \ldots & I_j(SC_1) & \ldots & I_n(SC_1) \\ \ldots & & \ldots & & \ldots \\ I_1(SC_i) & \ldots & I_j(SC_i) & \ldots & I_n(SC_i) \\ \ldots & & \ldots & & \ldots \\ I_1(SC_m) & \ldots & I_j(SC_m) & \ldots & I_n(SC_m) \end{Bmatrix}$$

Figure 4.   A decision matrix D for multi-criteria decision-making analysis, in which I denotes the indicator and SC denotes the scenario.

(qualitative versus quantitative), and the degree of uncertainty. Lin *et al.* [61] provided an extensive overview of different multi-criteria decision-making methods used in combination with LCSA.

A challenge for the actual execution of an ex-ante assessment is the uncertainty that is present in the data used to quantify the indicators when assessing new technologies, products, and their value chains. To take into account not only the uncertainty of data [62] but also the uncertainty regarding human judgments to determine weights [24], MCDA can involve interval numbers, fuzzy numbers, and rough numbers [61]. Stochastic multi-attribute analysis (SMAA) is an outranking method, based on PROMETHEE, which can be applied to compare different scenarios (e.g. different technology options) based on a multitude of sustainability indicators [63]. Stochasticity of data can be included by repeated random sampling of a set of input parameter values, which follow appropriate distributions [9]. Depending on the choices regarding the evaluation parameters in the SMAA model, the enforcement of a weak sustainability view can be restricted [9]. An illustrative example of an SMAA application and result is shown in Figure 5. SMAA offers a relative comparison of the sustainability of a multitude of technology or product scenarios, considering data uncertainty. It gives information on the most and least sustainable

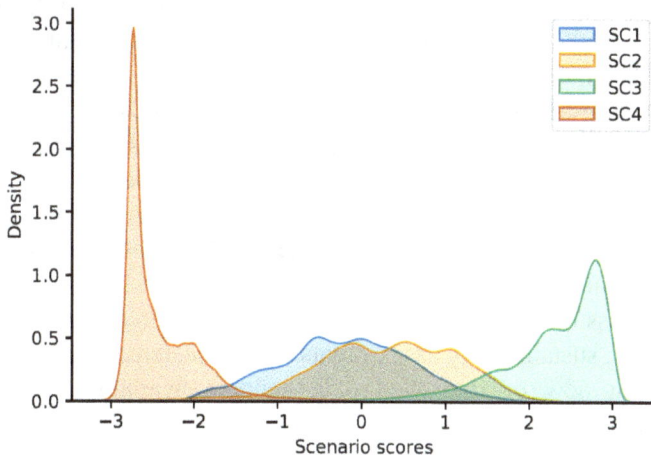

Figure 5. An illustrative example of an SMAA result comparing four scenarios on their sustainability. The illustrative data is visualized by a kernel density estimation. The higher the score, the better the sustainability performance.

alternatives, given the assumptions made during the full assessment process. The general aim is to further guide decision makers in research and development to make sustainable (investment) decisions.

However, by applying MCDA and integrating toward scenario scores, as is shown in Figure 5, the risk exists to lose information on the individual environmental, social, and economic indicators that were calculated in the previous TSA steps. It is therefore recommended to draw conclusions based on the individual indicator results, supplemented with an MCDA approach, which allows taking into account uncertainty while making early-stage decisions. For example, in Figure 5, a decision maker can decide to exclude scenario 4, as it is defined as the least sustainable scenario compared to the others. However, based on the results of the sensitivity analysis, it can also be decided to tackle the current bottlenecks and try to make changes toward the technological parameters of the assessed scenario.

The TSA results can be interpreted in a relative or absolute way. The SMAA method is focused on a relative comparison of the results between different scenarios. This is especially valuable at low TRL when certain technological and product-specific choices need to be made to select the most sustainable alternative. However, a product or technology can be more environmentally, economically, and socially sustainable than its benchmark, but the absolute sustainability performance can still be poor. Absolute sustainability assessment might raise questions on "should we even use or provide this product or service"? Chandrakumar *et al.* [64] introduced an absolute sustainability-based life cycle assessment (ASLCA). They focused on the climate impacts of different agri-food systems in absolute terms. ASLCA benchmarks the sustainability impacts toward sectoral impacts. However, interpreting the absolute results can be complex and often misleading because they depend on the delineation of the system boundaries and the level of detail of the data used to measure the sustainability impacts.

Regardless of whether the TSA results are interpreted in a relative or absolute way, sustainability assessments should be iterative along the TRL scale. When new or improved data become available, technologies mature, or new alternative scenarios are introduced based on previous results, a new "follow-up" TSA should be conducted. The two returning arrows in the integrated TSA figure (see Figure 2) represent these feedback loops to improve the sustainability assessments over time.

6.2. *Accounting for time-varying impacts*

To account for impact dynamics and managerial flexibility in a techno-economic context, the real options approach is considered to be more appropriate than cost−benefit or net present value analyses. The real options approach recognizes that costs and benefits are not constant but fluctuate in time, as these are subject to unexpected events and outcomes as well as learning and fluctuating price processes. It is recognized that the flexibility of waiting with an investment and the flexibility to adopt adaptive measures have value and should be integrated into decision support methods [65]. Flexibility options include the option to delay an investment, the option to develop in multiple stages, the option to expand or contract, the option to abandon a project, the option to temporarily suspend operations, the option to switch inputs or outputs, and growth options [66].

The real options approach is dynamic as it takes into account uncertainty with the stochastic evolution of costs and benefits through time. The inclusion of a time component in analyses that evaluate the feasibility of technologies is highly relevant when the adoption of these technologies is considered crucial to reach environmental policy objectives within specific long- and short-term time frames. Different quantitative techniques can be used to value these flexibility options and to determine an optimal time to invest [65, 67, 68].

However, in most conventional LCA or LCSA applications, various companies are incorporating environmental cost information into their accounting systems to prioritize investments in new technologies and products, without considering that these impacts may also vary in time [69]. To overcome this challenge and provide a solution forward, Compernolle *et al.* [70], for instance, combine Monte Carlo techniques with a decision tree. Monte Carlo simulations are run to evaluate the state-of-the-world of the stochastic processes at yearly points in time to select amongst the options available the one with the lowest risk−reward ratio. This method gives insights into the technical and economic boundary conditions that determine the selection of a specific technology development pathway and the associated benefit−risk balance. Although Compernolle *et al.* [70] do not include environmental impacts in their assessment, their methodology could be extended by including LCA results as well. Also, Kern *et al.* [71] propose to combine LCA and TEA in a real options framework to inform the design of algal biofuel production facilities [71]. In the LCA–TEA model, different plant designs under a range of potential future price scenarios were simulated. Hence, to evaluate the sustainability of technological developments through time, a methodological framework is needed that on the one hand accounts for a plurality of values and that on the other hand recognizes that technological innovations are made in consecutive steps to deal with uncertainty, acknowledging managerial flexibility and adaptations through time. Future research should consider developing a practical approach where the proposed integrated TSA framework is combined with real options-based decision trees such that the optimal time to exercise a specific technology development option will be defined based on multiple criteria instead of a single economic value.

7. Conclusion

The sustainability of new technologies and products should be assessed and monitored during their development. An integrated techno-sustainability assessment (TSA) is proposed to deal with the challenges regarding early development sustainability evaluations. An integrated TSA includes the following crucial steps and features:

- The alignment on system boundaries and the functional unit within the assessment, and thus, for all sustainability dimensions included.
- An extensive indicator selection, adjusted to case-specific needs. To measure sustainability as a whole, indicators should be selected from multiple disciplines.
- A dynamic quantification of the selected indicators, in which technological parameters are linked to sustainability indicators as much as possible.
- Uncertainties should be considered related to the data (collection), the market, and the importance of the selected indicators.
- TSA should be considered as an iterative process, where new data and movement along the TRL scale require new analyses.

The concept of sustainability and what it should entail will keep on changing over time depending on new insights and information. Future research will continue to improve current assessment methods for all sustainability dimensions and build new databases to fill out data gaps.

References

[1] International Union for Conservation of Nature and Natural Resources, *World Conservation Strategy: Living Resource Conservation for Sustainable Development* (Gland, Switzerland, 1980).

[2] W. Kloepffer, Life cycle sustainability assessment of products. *Int. J. Life Cycle Assess.* **13**, 89–95 (2008).

[3] UNEP (UN Environmental Programme), Towards a life cycle sustainability assessment (2011).

[4] R. T. Fauzi, P. Lavoie, L. Sorelli, M. D. Heidari, and B. Amor, Exploring the current challenges and opportunities of life cycle sustainability assessment. *Sustainability* **11**, 636 (2019).

[5] J. Guinée, Life cycle sustainability assessment: What is it and what are its challenges? In: R. Clift and A. Druckman (Eds.), *Taking Stock of Industrial Ecology* (Springer, Cham, 2016).

[6] R. Hoogmartens, S. Van Passel, K. Van Acker, and M. Dubois, Bridging the gap between LCA, LCC and CBA as sustainability assessment tools. *Environ. Impact Assess. Rev.* **48**, 27–33 (2014).

[7] K. Raworth, Doughnut economics: 7 ways to think like a 21st-century economist (2017).

[8] C. Wulf, J. Werker, P. Zapp, A. Schreiber, H. Schlör, and W. Kuckshinrichs, Sustainable development goals as a guideline for indicator selection in life cycle sustainability assessment. *Procedia CIRP Life Cycle Eng. Conf.* **69**, 59–65 (2018).

[9] S. Van Schoubroeck, G. Thomassen, S. Van Passel, R. Malina, J. Springael, S. Lizin *et al.*, An integrated techno-sustainability assessment (TSA) framework for emerging technologies. *Green Chem.* **23**, 1700 (2021).

[10] A. Gasparatos and A. Scolobig, Choosing the most appropriate sustainability assessment tool. *Ecol. Econ.* **80**, 1–7 (2012).

[11] I. Juwana, N. Muttil, and B. J. C. Perera, Indicator-based water sustainability assessment — A review. *Sci. Total Environ.* **438**, 357–371 (2012).

[12] M. Lundin, Indicators for measuring the sustainability of urban water systems — A life cycle approach (2003).

[13] S. Van Schoubroeck, M. Van Dael, S. Van Passel, and R. Malina, A review of sustainability indicators for biobased chemicals. *Renew. Sustain. Energy Rev.* **94**, 115–126 (2018).

[14] C. Schader, J. Grenz, M. S. Meier, and M. Stolze, Scope and precision of sustainability assessment approaches to food systems. *Ecol. Soc.* **19** (2014). https://www.ecologyandsociety.org/vol19/iss3/art42/

[15] A. Mascarenhas, L. M. Nunes, and T. B. Ramos, Selection of sustainability indicators for planning: Combining stakeholders' participation and data reduction techniques. *J. Clean. Prod.* **92**, 295–307 (2015).

[16] C. Bockstaller and P. Girardin, How to validate environmental indicators. *Agric. Syst.* **76**, 639–653 (2003).

[17] T. Kurka and D. Blackwood, Participatory selection of sustainability criteria and indicators for bioenergy developments. *Renew. Sustain. Energy Rev.* **24**, 92–102 (2013).

[18] N. Dalkey and O. Helmer, An experimental application of the DELPHI method to the use of experts. (1962).

[19] C. Okoli and S. D. Pawlowski, The Delphi method as a research tool: An example, design considerations and applications. *Inf. Manag.* **42**, 15–29 (2004).

[20] S. Van Schoubroeck, J. Springael, M. Van Dael, R. Malina, and S. Van Passel, Sustainability indicators for biobased chemicals: A Delphi study using Multi-Criteria Decision Analysis. *Resour. Conserv. Recycl.* **144**, 198–208 (2019).

[21] M. G. Kendall, A new measure of rank correlation. *Biometrika* **30**, 81–89 (1938).

[22] C. Spearman, The proof and measurement of association between two things. *Am. J. Psychol.* **15**, 72 (1904).

[23] W. De Keyser and J. Springael, *Why Don't We Kiss?* (University Press Antwerp, Belgium, 2009).

[24] M. Bengtsson and B. Steen, Weighting in LCA — Approaches and applications. *Environ. Progress* **19**, 101–109 (2000).

[25] J. C. Mankins, Technology readiness assessments: A retrospective. *Acta Astronaut.* **65**, 1216–1223 (2009).

[26] G. Thomassen, M. Van Dael, S. Van Passel, and F. You, How to assess the potential of emerging green technologies? Towards a prospective environmental and techno-economic assessment framework. *Green Chem.* **21**, 4868–4886 (2019).

[27] F. Piccinno, R. Hischier, S. Seeger, and C. Som, From laboratory to industrial scale: A scale-up framework for chemical processes in life cycle assessment studies. *J. Clean. Prod.* **135**, 1085–1097 (2016).

[28] M. Buyle, A. Audenaert, P. Billen, K. Boonen, and S. Van Passel, The future of Ex-Ante LCA? Lessons learned and practical recommendations. *Sustainability* **11**, 5456 (2019).

[29] R. Arvidsson, A. Tillman, B. A. Sandén, M. Janssen, A. Nordelöf, D. Kushnir *et al.*, Environmental assessment of emerging technologies: Recommendations for prospective LCA. *J. Ind. Ecol.* **22**, 1286–1294 (2018).

[30] M. K. Hulst, M. A. J. Huijbregts, N. Loon, M. Theelen, L. Kootstra, J. D. Bergesen *et al.*, A systematic approach to assess the environmental impact of emerging technologies: A case study for the GHG footprint of CIGS solar photovoltaic laminate. *J. Ind. Ecol.* **24**, 1234–1249 (2020).

[31] T. Kuppens, M. Van Dael, K. Vanreppelen, T. Thewys, J. Yperman, R. Carleer *et al.*, Techno-economic assessment of fast pyrolysis for the valorization of short rotation coppice cultivated for phytoextraction. *J. Clean. Prod.* **88**, 336–344 (2015).

[32] J. Wunderlich, K. Armstrong, G. A. Buchner, P. Styring, and R. Schomäcker, Integration of techno-economic and life cycle assessment: Defining and applying integration types for chemical technology development. *J. Clean. Prod.* **287**, 125021 (2020).

[33] A. W. Zimmermann, J. Wunderlich, L. Müller, G. A. Buchner, A. Marxen, S. Michailos *et al.*, Techno-economic assessment guidelines for CO_2 utilization. *Front. Energy Res.* **8** (2020). https://www.frontiersin.org/articles/ 10.3389/fenrg.2020.00005/full

[34] K. Roh, A. Bardow, D. Bongartz, J. Burre, W. Chung, S. Deutz *et al.*, Early-stage evaluation of emerging CO2 utilization technologies at low technology readiness levels. *Green Chem.* **22**, 3842–3859 (2020).

[35] G. Centi, S. Perathoner, A. Salladini, and G. Iaquaniello, Economics of CO_2 utilization: A critical analysis. *Front. Energy Res.* **8** (2020). https://www. frontiersin.org/articles/10.3389/fenrg.2020.567986/full

[36] M. Van Dael, T. Kuppens, S. Lizin, and S. Van Passel, Techno-economic assessment methodology for ultrasonic production of biofuels. In: Z. Fang, J. R. Smith, and X. Qi (Eds.), *Production of Biofuels and Chemicals with Ultrasound* (Springer, Dordrecht, 2015).

[37] M. Larrain, S. Van Passel, G. Thomassen, B. Van Gorp, T. T. Nhu, S. Huysveld *et al.*, Techno-economic assessment of mechanical recycling of challenging post-consumer plastic packaging waste. *Res. Cons. Recycl.* **170**, 105607 (2021).

[38] M. Tschulkow, T. Compernolle, S. Van den Bosch, J. Van Aelst, I. Storms, M. Van Dael *et al.*, Integrated techno-economic assessment of a biorefinery process: The high-end valorization of the lignocellulosic fraction in wood streams. *J. Clean. Prod.* **266**, 122022 (2020).

[39] G. Thomassen, M. Van Dael, and S. Van Passel, The potential of microalgae biorefineries in Belgium and India: An environmental techno-economic assessment. *Bioresour. Technol.* **267**, 271–280 (2018).

[40] F. Vanclay, International principles for social impact assessment. *Impact Assess. Proj. Apprais.* **21**, 5–12 (2003).

[41] C. Benoît Norris, M. Traverso, S. Neugebauer, E. Ekener, T. Schaubroeck, S. Russo Garrido *et al.*, Guidelines for social life cycle assessment of products and organizations (2020).

[42] P. Rafiaani, T. Kuppens, M. Van Dael, H. Azadi, P. Lebailly, and S. Van Passel, Social sustainability assessments in the biobased economy: Towards a systemic approach. *Renew. Sustain. Energy Rev.* **82**, 1839–1853 (2018).

[43] C. B. Norris, Data for social LCA. *Int. J. Life Cycle Assess.* **19**, 261–265 (2014).

[44] O. Tokede and M. Traverso, Implementing the guidelines for social life cycle assessment: Past, present, and future. *Int. J. Life Cycle Assess.* **25**, 1910–1929 (2020).

[45] S. González-García, P. Gullón, and B. Gullón, Bio-compounds production from agri-food wastes under a biorefinery approach: Exploring environmental and social sustainability. In *Quantification of Sustainability Indicators in the Food Sector. Environmental Footprints and Eco-Design of Products and Processes* (Springer, Singapore, 2019).

[46] P. Rafiaani, T. Kuppens, G. Thomassen, M. Van Dael, H. Azadi, P. Lebailly *et al.*, A critical view on social performance assessment at company level: Social life cycle analysis of an algae case. *Int. J. Life Cycle Assess.* **25**, 363–381 (2020).

[47] A. W. Zimmermann and R. Schomäcker, Assessing early-stage CO_2 utilization technologies-comparing apples and oranges? *Energy Technol.* **5**, 850–860 (2017).

[48] S. McCord, K. Armstrong, and P. Styring, Developing a triple helix approach for CO2 utilisation assessment. *Faraday Discuss.* **230**, 247–270 (2021).

[49] M.-O. P. Fortier, L. Teron, T. G. Reames, D. T. Munardy, and B. M. Sullivan, Introduction to evaluating energy justice across the life cycle: A social life cycle assessment approach. *Appl. Energy.* **236**, 211–219 (2019).

[50] I. Oosterlaken, Applying Value Sensitive Design (VSD) to wind turbines and wind parks: An exploration. *Sci. Eng. Ethics.* **21**, 359–379 (2015).

[51] R. Brouwer and L. R. van Ek, Integrated ecological, economic and social impact assessment of alternative flood control policies in the Netherlands. *Ecol. Econ.* **50**, 1–21 (2004).

[52] N. E. Matthews, L. Stamford, and P. Shapira, Aligning sustainability assessment with responsible research and innovation: Towards a framework for constructive sustainability assessment. *Sustain. Prod. Consum.* **20**, 58–73 (2019).

[53] K. Winans, F. Dlott, E. Harris, and J. Dlott, Sustainable value mapping and analysis methodology: Enabling stakeholder participation to develop localized indicators mapped to broader sustainable development goals. *J. Clean. Prod.* **291**, 125797 (2021).

[54] C. A. Frate, C. Brannstrom, M. V. G. de Morais, and A. de A. Caldeira-Pires, Procedural and distributive justice inform subjectivity regarding wind power: A case from Rio Grande do Norte, Brazil. *Energy Policy.* **132**, 185–195 (2019).

[55] A. Revez, N. Dunphy, C. Harris, G. Mullally, B. Lennon, and C. Gaffney, Beyond forecasting: Using a modified Delphi method to build upon participatory action research in developing principles for a just and inclusive energy transition. *Int. J. Qual. Methods.* **19**, 1–12 (2020).

[56] A. Tabi and R. Wüstenhagen, Keep it local and fish-friendly: Social acceptance of hydropower projects in Switzerland. *Renew. Sustain. Energy Rev.* **68**, 763–773 (2017).

[57] M. Van Der Spek, T. Fout, M. Garcia, V. Nair, M. Matuszewski, S. Mccoy *et al.*, Uncertainty analysis in the techno-economic assessment of CO_2 capture and storage technologies. Critical review and guidelines for use. *Int. J. Greenh. Gas Control.* **100**, 103113 (2020).

[58] R. Brun, M. Kühni, H. Siegrist, W. Gujer, and P. Reichert, Practical identifiability of ASM2d parameters — Systematic selection and tuning of parameter subsets. *Water Res.* **36**, 4113–4127 (2002).

[59] A. Ciroth, S. Muller, B. Weidema, and P. Lesage, Empirically based uncertainty factors for the pedigree matrix in ecoinvent. *Int. J. Life Cycle Assess.* **21**, 1338–1348 (2016).

[60] G. Thomassen, S. Van Passel, and J. Dewulf, A review on learning effects in prospective technology assessment. *Renew. Sustain. Energy Rev.* **130**, 109937 (2020).

[61] R. Lin, Y. Man, and J. Ren, Life cycle decision support framework: Method and case study. In Jingzheng Ren and Sara Toniolo (Eds.), *Life Cycle Sustainability Assessment for Decision-Making* (Elsevier Inc., 2020).

[62] H. H. Khoo, V. Isoni, and P. N. Sharratt, LCI data selection criteria for a multidisciplinary research team: LCA applied to solvents and chemicals. *Sustain. Prod. Consump.* **16**, 68–87 (2018).

[63] V. Prado and R. Heijungs, Implementation of stochastic multi attribute analysis (SMAA) in comparative environmental assessments. *Environ. Model. Softw.* **109**, 223–231 (2018).

[64] C. Chandrakumar, S. J. McLaren, N. P. Jayamaha, and T. Ramilan, Absolute Sustainability-Based Life Cycle Assessment (ASLCA): A benchmarking approach to operate agri-food systems within the 2°C global carbon budget. *J. Ind. Ecol.* **23**, 906–917 (2019).

[65] A. K. Dixit and R. S. Pindyck, Investment under uncertainty. (2012).

[66] L. Trigeorgis, *Real Options: Managerial Flexibility and Strategy in Resource Allocation* (MIT Press, Cambridge, 2004).

[67] K. J. M. Huisman and P. M. Kort, Strategic capacity investment under uncertainty. *RAND J. Econ.* **46**, 376–408 (2015).

[68] T. Dangl, Investment and capacity choice under uncertain demand. *Eur. J. Oper. Res.* **117**, 415–428 (1999).

[69] K. G. Shapiro, Incorporating costs in LCA. *Int. J. LCA* **6**, 121–123 (2001).

[70] T. Compernolle, K. Welkenhuysen, E. Petitclerc, D. Maes, and K. Piessens, The impact of policy measures on profitability and risk in geothermal energy investments. *Energy Econ.* **84**, 104524 (2019).

[71] J. D. Kern, A. M. Hise, G. W. Characklis, R. Gerlach, S. Viamajala, and R. D. Gardner, Using life cycle assessment and techno-economic analysis in a real options framework to inform the design of algal biofuel production facilities. *Bioresour. Technol.* **225**, 418–428 (2017).

Chapter 15

LCA Applications for Materials Recovery from Waste-to-Energy Technologies: Concepts from the Three Pillars of Sustainability

Gamini P. MENDIS[*,‡] **and Justin S. RICHTER**[†,§]

Plastics Engineering Technology, The Pennsylvania State University, The Behrend Campus, 5101 Jordan Road, Erie, PA 16508, USA

†*Agricultural and Biological Engineering, The Pennsylvania State University, 105 Agricultural Engineering Building, Shortlidge Road, University Park, PA 16802, USA*

‡*gmendis@psu.edu*

§*jsrichter@psu.edu*

Waste-to-energy (WTE) technologies are widely used across the world to reduce the volume of waste entering landfills and recover energy. New technologies are emerging to recover additional value from WTE residue. LCA research in the area of WTE has recently expanded beyond technical feasibility and environmental impacts to include issues of equity, justice, social impacts, and stakeholder engagement. This chapter discusses new advances in WTE technologies and corresponding advances in life cycle assessment. The three pillars of sustainability (economy, environment, and society) methods for LCA in waste management systems are also presented.

1. Introduction

According to *World Bank* statistics, global waste generation is expected to increase from 2.01 billion metric tons in 2020 to 3.40 billion metric tons in 2050 [1]. As the global population grows, accompanied with the development of industrial nations, societies are expected to become more affluent [2]. In ideal circumstances, waste is managed according to the waste management hierarchy (Figure 1), where reducing, reusing, and recycling waste are more favored than landfilling and energy recovery. However, in many circumstances, source separation of waste materials is not possible for economic or logistics reasons. In these scenarios, waste-to-energy (WTE) facilities can be used to reduce the volume of waste being sent to landfills and recover energy and/or materials from municipal solid waste (MSW). Over 2,450 traditional WTE facilities are in operation globally [3], with most new facilities being built in China and the EU [4], while a significant number of existing facilities are operational in Japan [5, 6].

In terms of research methods, life cycle assessment is the most popular method. The main research fields focus on environmental and social

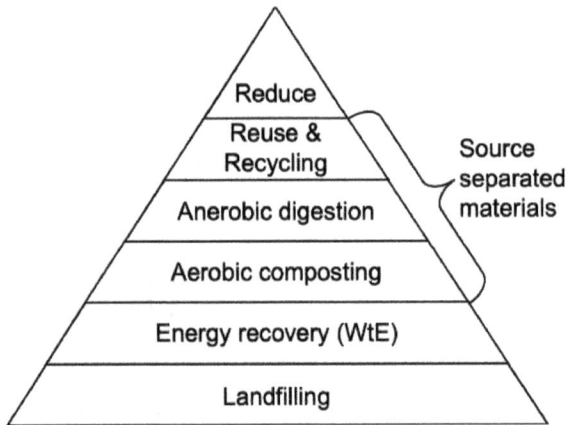

Figure 1.   The hierarchy of waste management options.

impacts, and energy technology innovations, and valuable product recovery.

1.1. *International considerations of waste and residues*

Current international waste regulations (e.g. Renewable Energy Directive (RED EU)) define waste and residues (W&R) from both agricultural and industrial processes as materials not intended for end-product consumption [7, 8]. W&R effectively have zero emissions because according to international guidelines, all process emissions are to be allocated to any co-products. This results in waste products being reduced to a zero emissions feedstock. The zero emissions assumption can incentivize repurposing, recycling, or reuse (Figure 1) of W&R, thus removing W&R out of the WTE stream. An increase in recycling or reuse schemes aims to reduce the amount of waste available for a WTE facility. In such cases, W&R resources generated from WTE may decrease. Increasingly, international conventions, regulations, and guidelines are incentivizing W&R recapture and utilization within the same facility, e.g. syngas from waste plastic pyrolysis; this simultaneously reduces process waste and offsets fossil fuel energy consumed for any main products. At present, technology is the limiting factor to recover value from waste streams, but industry interest in closing material loops and use of WTE technologies is expanding [9, 10].

1.2. *General waste-to-energy facility operations*

WTE facilities can take a variety of forms. The most common type of WTE facility is an incineration or combustion facility, which converts waste directly into heat, which is then used to power steam turbine generators. In some instances, WTE facilities are also combined heat and power (CHP) complexes that use the heat from the steam in industrial or household heating systems. The Amager Bakke plant in Copenhagen is one example of such a facility which uses waste to heat residential housing throughout Copenhagen.

Several steps are required to convert waste into heat and/or energy; a thorough discussion can be found in the book by Themelis and Bourtsalas [11]. To summarize, waste is received, non-combustible fractions are removed, and the waste is mixed in a pit to homogenize the composition of the waste. Modern technologies are exploring the valorization of WTE by-products to create new value and offset operating costs. Emerging innovations that are set to recover valuable products from waste fractions will be helpful in reducing waste volumes as well as beneficial in reducing environmental pollution.

1.3. *Value recovery from waste-to-energy systems*

WTE facilities recover value from waste by extracting energy or materials. Energy is recovered by extracting electricity from steam or using the heat from steam. The heat created from incineration is applied to steam boilers that then generate electricity through turbine generators. It is worthwhile to consider that if future electricity prices drop (e.g. due to competitive renewable facilities), this step may become less economically viable. In some facilities, the steam is also distributed for industrial or residential heating applications. Energy from steam boilers can supplement the consumption of traditional fuels such as natural gas. Co-location of WTE facilities with other energy-intensive industries (e.g. cement production, steel manufacturing) may be advantageous. While heat and energy are the most common main products, other material products from incineration are proving to be valuable resources as well [12].

A myriad of materials can be recovered from waste. Conventional techniques use magnets and eddy current separators to remove ferrous and non-ferrous metals, respectively. In some newer systems, such as the Ash Recycling and Processing Facility at York County Solid Waste Authority in Pennsylvania, USA, aggregate, char, metals, and mineral-like fractions are also cleaned and/or separated from bottom ash [13]; in these systems, vibrating grates, presses, magnetic separators, and sieves are used in combination to reduce the size of the ash and separate it into valorizable fractions. The metals, aggregate, and mineral fractions can be sold to recover value from the waste [14]. Unused fractions of the bottom ash and fly ash are landfilled, however, up to 90% of the waste volume is reduced through the incineration process [13], saving landfill space and reducing tipping fees for the WTE facility.

The landfilled ash fractions may still contain valuable material, as evidenced by the emerging attention to rare earth element (REE) recovery from ashes. REEs, such as neodymium, cerium, and yttrium, are used in many advanced technologies, but they are difficult and costly to extract with considerable supply chain challenges [15, 16]. There is potential for the concentration of REEs in ash to be higher than that found in mineral sources [17], which may encourage new technology development to recover REEs. However, the chemical complexity of bottom ash may be prohibitive and the highest concentration of REEs tends to be in consumer electronics; combustion of electronics has been shown to create toxic emissions [18] (e.g. cadmium, lead, dioxins). Since electronic waste (e-waste) is one of the fastest growing pollution problems worldwide, leading to continuous contamination of ecosystems, enhanced management of e-waste will tend to benefit both the environment and social health issues.

Due to the complexity and variety of WTE processing technologies and the potential for hazardous emissions, it is important to conduct an LCA on WTE systems. LCAs are used to identify opportunities for system improvement and reduction of overall environmental and social impacts. The following section describes pertinent aspects, specific challenges, and new directions of an LCA for WTE systems.

2. Examples of Waste-to-Energy Case Studies

Several LCA studies have been performed on WTE systems with varying degrees of comprehensiveness and scope of investigation [18, 19]. Consistent with all LCAs is the requirement to define the goal, scope, functional unit, and describe the system boundaries. While the goal and scope for WTE systems may be the same/similar, the functional unit can vary. Functional units are most often selected based on mass of waste [20]; the amount of waste generated in a geographic region or the amount of waste entering or leaving a treatment facility are all appropriate functional units.

Further complications in comparative LCA investigations are the selection of activities or stages to be included in system boundaries (refer to Chapter 1 for details). Like other systems evaluated with LCA, the selection of a system boundary for WTE systems is required. The following steps are used for identifying the system boundary for a WTE operation: (1) point of origin or source of waste, (2) transportation to WTE facility, (3) the facility, its technology, and efficiency, and (4) alternative disposal baselines (including recycling).

The boundary for a WTE system generally requires that waste types and amounts and their composition be included from the point of origin to the incineration process. Generally, when evaluating system boundaries for a WTE operation, the waste feedstocks are sourced from external processes, e.g. plastic waste collected from the curbside, a recycler/sorter, or a Materials Recycling Facility (MRF). The type of waste also has a significant influence on the total environmental impacts of the process and must be carefully defined. In any case, due to the nature of waste as an undesirable result of production, the source of waste is often the starting point for a system boundary.

Transportation of waste, both upstream and downstream of the WTE facility, must also be considered when drawing a system boundary (see Figure 2). The point of origin ("cradle" in the WTE scenario) for waste is often the end of life ("grave") for a product or material. In various LCA case studies, logistical arrangements are also considered for the waste generated at one facility or household to a WTE processing facility. Collection and transportation to the WTE facility are important considerations especially when alternatives (e.g. landfills) are often located at a significant distance away from the point of collection. In many urban locations, garbage trucks and semi-trucks may all be used to transport waste to transfer stations, the WTE facility, recycling locations, or landfill sites. Due to different waste collection locations and distances, the logistics involved in collection and disposal are often very complex and can be challenging to model and validate in LCA studies.

After the waste is collected and transported to the WTE facility, it will be sent for the next stage or level of processing. The operating conditions or parameters of the recycling facility can greatly affect the environmental impacts of the process. The type of thermal treatment, the type of reactor, and the set of air pollution control technologies (dust control, acid gas control, dioxin/furan removal, SO_x/NO_x removal) are all important process parameters which can considerably impact the heat and energy recovery and the emissions from the process [20, 21].

Historically, WTE technologies have been considered "dirty" technologies. The emission of dioxins, furans, heavy metals, SO_x, NO_x, and ash has all previously been problematic [22]; due to such concerns, new advancements in WTE designs and recycling systems will gradually play a significant role. Over the last 30 years, new emission control technologies and regulations have greatly reduced the environmental impact of combustion/incineration facilities. The technology that the WTE facility utilizes will determine if additional by-products are created (Figure 2), e.g. bottom ash, fly ash, sludge, metals, aggregate, or recoverable gas.

In many waste management systems, it is important to address the avoided burdens by including alternative disposal baselines in the system

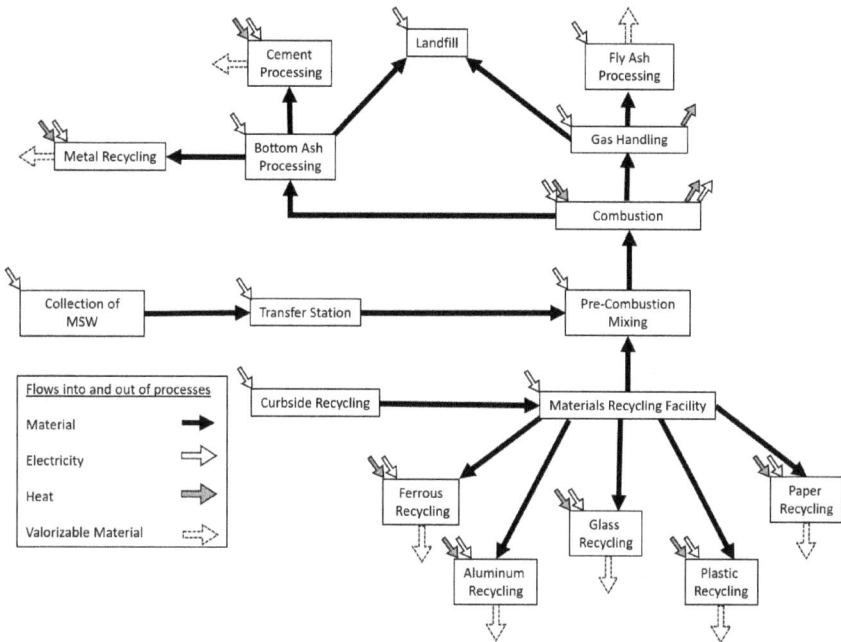

Figure 2.   Examples of system boundaries and material and energy flows from a waste collection system.

boundaries. When energy, heat, metals, or aggregate are recovered from the WTE process, the avoided impacts of the primary production of those by-products can be credited to the system. If the WTE process involves a significant reduction in the amount of waste sent to landfills, then the associated landfill leachate [23–25] and fugitive methane emissions [26, 27] can be avoided. Unfortunately, conventional mass burn systems for municipal solid waste may emit large quantities of acidic gases, dioxins, as well as heavy metals. Toxic emissions affect all the people on the planet, therefore, LCA impacts on the stakeholders have recently been considered.

3. Emerging Developments in Life Cycle Assessment of Waste-to-Energy Systems

WTE systems are not new in the scheme of energy production, but LCA research in the area of WTE has recently expanded beyond technical feasibility and environmental impacts to include issues of equity, justice, social impacts, and stakeholder engagement. The following section discusses the social impacts of WTE systems, followed by a description of uncertainties and model improvements in comparative WTE LCAs, while finally presenting a brief exploration of LCA tools for WTE technologies.

3.1. *Stakeholder and social impacts*

With growing attention to social impacts from production, integrating social LCA (S-LCA) into the evaluation of a WTE facility is becoming a mandatory practice. A social LCA can identify negative social impacts from the WTE supply chain while also suggesting opportunities for intervention. However, limited studies are available currently that apply S-LCA principles to WTE technologies. Where WTE facilities exist, stakeholders can have varying degrees of interest and requirements. The most visible stakeholders of a WTE facility can range from the workers/employees and shareholders to the energy consumers and the local community (see Figure 3) [28, 29]. A plethora of indirect stakeholders may also be impacted by the operations of a WTE facility and these may be much more difficult to identify and quantify. Further complicating the

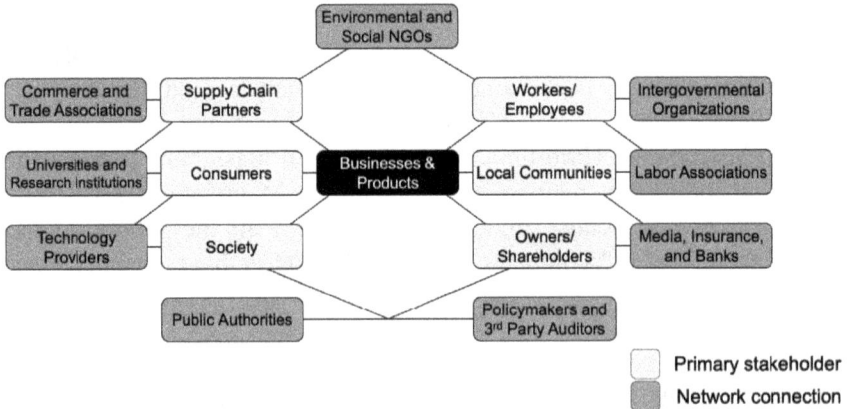

Figure 3.   Spoke and wheel diagram of common stakeholders (direct primary and potential indirect network connections) for businesses and products.

Source: Adapted from Ref. [29].

quantification challenge is a lack of suitable measures, metrics, or indicators [29] that effectively describe the social impacts of WTE systems.

3.1.1. *Advances in quantifying social impacts*

Zhou *et al.* [30] accurately identify that S-LCA methodologies have not been applied to assess WTE technologies and few S-LCA studies evaluate waste treatment scenarios. Further, S-LCA studies are often non-comparable with vastly differing functional units suggesting that there is no single agreed-upon unit of social impact applicable to all scenarios and across various scales [31]. The current best practices for evaluating social impacts include assessing pollution and emissions (an environmental concern) as a dollar cost per ton [32, 33], frequently referencing the effect on human health. However, such analysis is limited, as the variety of impacted stakeholders are not solely individuals but may consist of groups with very different needs. It becomes evident that any industry seeking to internalize the true social impacts created by industrial activity must attempt to quantify the impacts on each stakeholder, for which no model, tool, software, or dataset yet provides.

There has been progress in reports and literature where various products and services have been analyzed for social impacts [34] following the United Nations Environment Programme (UNEP) Guidelines for S-LCA [35, 36]. These studies vary in the application of the myriad qualitative and quantitative indicators suggested by Benoît-Norris *et al.* [37]. Rasmussen *et al.* [38] found that the number of social indicators across the literature is expanding, suggesting that convergence on scalable and suitable indicators is still in the exploratory stage of research.

One approach suggested a collection of statistically "best fit" social indicators [39] that address stakeholder needs. A different approach proposed a collection of sustainability indicators (Figure 4) focused on WTE systems and linked the three pillars of sustainability (economy, environment, and society) into a scaled metric of sustainability [40]. While the area of research is moving closer toward specific and applicable social impact indicators, the vast majority of social indicators lack the level of rich data that has populated environmental LCA studies for decades.

Where quantified indicators have been applied, these are typically analyzed using extended economic input–output (IO) methodologies (for a comprehensive IO explanation, see Miller and Blair [41]) [42–44]. The IO method can capably identify industry- and region-wide impacts in

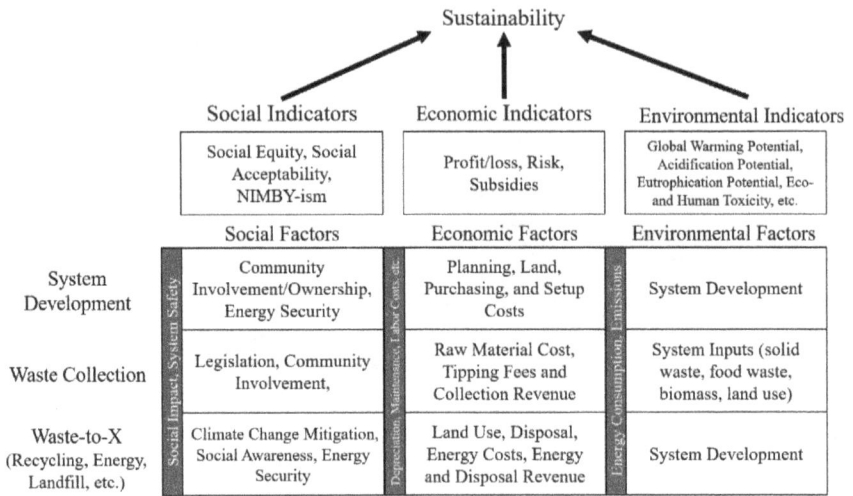

Figure 4.   Proposed collection of indicators for the three dimensions of sustainability (economy, environment, and society) applied to WTE systems.

Source: Adapted from Ref. [40].

supply chains for products, services, and industries. However, at present, in the absence of true causal models for social impacts of production activity, IO models are the best available option for quantifying social impacts.

3.1.2. *Equity, justice, and stakeholder voice*

Social and environmental justice may be of further concern during the siting, construction, operation, and decommissioning of a WTE facility [45]. A common practice during WTE siting (particularly in the U.S.) is to evaluate the land near economically depressed and politically underrepresented communities. In these locations, there may be limited objection to economic development, insufficient organization to oppose development, or further limited understanding of the entire life cycle impacts of a new WTE facility. The long-term and human health impacts of WTE generation facilities are also not fully understood [46, 47]; however, for a specific case study in China, Yuen *et al.* [48] reported that environmental impacts from WTE are relatively lower than traditional landfills. In other (more general) cases, there is often significant negative public perception about WTE or incineration facilities [49–51].

3.1.3. *Avoided impacts from replacement technologies*

Apart from negative environmental impacts, the benefits of a WTE facility can be categorized by the avoided emissions that otherwise would be produced from a traditional energy generation facility (e.g. coal-fired power plants), a traditional landfill, or from open burning of waste. A reduced or avoided impact relative to the *status quo* can also be a positive impact in life cycle terminology. For example, some modern conversion technologies or gasification systems are capable of converting used cooking oil into bio-derived energy or fuels [52] and can effectively result in the avoidance of emissions from conventionally produced fossil fuels. The social and environmental benefits of WTE facilities can, therefore, be considered a function of the alternative technologies available in a location. LCA can identify the overall sustainability of a WTE system, but uncertainties remain.

3.2. *Uncertainties/opportunities for model improvement*

When performing a comparative LCA of various waste scenarios, it is imperative that appropriate baseline cases are selected. The most common comparison scenarios for WTE systems are landfilling and/or recycling. However, recent research has shown that previous assumptions about landfill gas emissions may be drastically underestimating emissions from landfill point sources [53] and that 10% of point sources can emit roughly 60% of emissions. Furthermore, recycling has been shown to have similar challenges with the accuracy of embedded assumptions, particularly with regard to the actual transportation distance of the material and the amount of material landfilled. These concerns should be carefully included in future LCA efforts to fully capture the environmental impacts of the system.

3.3. *Advancements in waste-to-energy and life cycle assessment tools*

Several WTE-specific LCA tools exist to help practitioners conduct analyses of new systems. In the US, the Municipal Solid Waste Decision Support Tool (MSWDST) and its successor, the Solid Waste Optimization Framework (SWOLF) [54] are available to analyze costs and environmental impacts of waste management decisions. In the EU, software such as EASETECH, ORWARE, and WRATE are available for project analysis [55, 56]. Some of these tools allow Monte Carlo simulations or sensitivity analysis, which are important techniques for evaluating and reducing uncertainty in the analysis.

Apart from LCA applications integrated with the three pillars of sustainability concepts (social, economy, environment) [28, 40, 49], it is also worth mentioning that new advances in emerging waste conversion technologies are ongoing. As one noteworthy example, Das *et al.* [57] presented a case for generating value-added products from the thermo-chemical treatment from e-waste. In another case, Gong *et al.* [58] converted plastic waste into high value-added carbon nanomaterials via a catalytic carbonization method. Interest in the conversion of waste into energy and valuable chemicals is also ongoing by applications of biological (e.g. anaerobic digestion) or thermochemical processes (e.g. pyrolysis) [59, 60].

4. Concluding Remarks

Waste conversion technologies such as WTE will continue to play a significant role in waste management and energy recovery sectors in the future. New technologies may emerge to capture considerable value from WTE by-products and interest in novel material recovery is growing globally. With the development of new technologies, it is important to understand the sustainability impacts of these WTE systems to identify circumstances where new technologies should or could be implemented. To conduct a thorough sustainability assessment or comparative analysis, carefully selected system boundaries and functional units must correspond with a suitable goal and scope. Future LCA studies should integrate the social, economic, and environmental impacts of WTE systems and should consider global and regional environmental regulations. Collectively, information created from a holistic LCA perspective will aid in making more sustainable decisions in the future.

References

[1]   S. Kaza, Y. C. Lisa, P. Bhada-Tata, and F. Van Woerden, *What a Waste 2.0: A Global Snapshot of Solid Waste Management to 2050* (World Bank, Washington, DC, 2018). https://datatopics.worldbank.org/what-a-waste/trends_in_solid_waste_management.html.

[2] M. J. Castaldi and N. J. Themelis, The case for increasing the global capacity for waste to energy (WTE). *Waste Biomass Valorization* **1**, 91–105 (2010).

[3] Ecoprog Gmbh, *Waste to Energy.* (2019). https://www.ecoprog.com/publikationen/energiewirtschaft/waste-to-energy/.

[4] D. Zhang, G. Huang, Y. Xu, and Q. Gong, Waste-to-energy in China: Key challenges and opportunities. *Energies* **8**, 14182–14196 (2015).

[5] N. J. Themelis and C. Mussche, Municipal solid waste management and waste-to-energy in the United States, China and Japan. *2nd Int. Acad. Symp. Enhanc. Landfill Min.* 1–19 (2013).

[6] L. Sun, Z. Li, M. Fujii, Y. Hijioka, and T. Fujita, Carbon footprint assessment for the waste management sector: A comparative analysis of China and Japan. *Front. Energ.* **12**, 400–410 (2018).

[7] European Parliament and Council, Directive (EU) 2018/2001 of the European Parliament and of the Council of 11 December 2018 on the promotion of the use of energy from renewable sources. Document 32018L2001 (2018).

[8] EUR-Lex, Commission notice on technical guidance on the classification of waste. Official J. of EU, C124/01 (2018).

[9] W. Haas, F. Krausmann, D. Wiedenhofer, and M. Heinz, How circular is the global economy? An assessment of material flows, waste production, and recycling in the European union and the world in 2005. *J. Ind. Ecol.* **19**, 765–777 (2015).

[10] J. N. Hahladakis, C. A. Velis, R. Werber, Costas, E. Iacovidou, and P. Purnell, An overview of chemical additives present in plastics: Migration, release, fate and environmental impact during their use, disposal and recycling. *J. Hazard. Mat.* **344**, 179–199 (2018).

[11] N. J. Themelis and A. C. Bourtsalas (Eds.), *Recovery of Materials and Energy from Urban Waste, A Volume in the Encyclopedia of Sustainability Science and Technology*, 2nd Edition (Springer, New York, US, 2019).

[12] J. Kim and S. Jeong, Economic and environmental cost analysis of incineration and recovery alternatives for flammable industrial waste: The case of South Korea. *Sustain.* **9**, 1638 (2017).

[13] A. M. Joseph, R. Snellings, P. Van den Heede, S. Matthys, and N. De Belie, The use of municipal solid waste incineration ash in various building materials: A Belgian point of view. *Mat.* **11**, 141 (2018).

[14] M. Šyc, F. G. Simon, J. Hykš, R. Braga, L. Biganzoli, G. Costa *et al.*, Metal recovery from incineration bottom ash: State-of-the-art and recent developments. *J. Hazard. Mat.* **393**, 122433 (2020).

[15] J. Burlakovs, Y. Jani, M. Kriipsalu, Z. Vincevica-Gaile, F. Kaczala, G. Celma *et al.*, On the way to "zero waste" management: Recovery potential of elements, including rare earth elements, from fine fraction of waste. *J. Clean. Prod.* **186**, 81–90 (2018).

[16] D. Bauer, D. Diamond, J. Li, D. Sandalow, P. Telleen, and B. Wanner, *Critical Materials Strategy* (U.S. Department of Energy, 2010). US.

[17] L. S. Morf, R. Gloor, O. Haag, M. Haupt, S. Skutan, F. Di Lorenzo *et al.*, Precious metals and rare earth elements in municipal solid waste — Sources and fate in a Swiss incineration plant. *Waste Manag.* **33**, 634–644 (2013).

[18] P. Kiddee, R. Naidu, and M. H. Wong, Electronic waste management approaches: An overview. *Waste Manag.* **33**, 1237–1250 (2013).

[19] F. Mayer, R. Bhandari, and S. Gäth, Critical review on life cycle assessment of conventional and innovative waste-to-energy technologies. *Sci. Total Environ.* **672**, 708–721 (2019).

[20] T. F. Astrup, D. Tonini, R. Turconi, and A. Boldrin, Life cycle assessment of thermal Waste-to-Energy technologies: Review and recommendations. *Waste Manag.* **37**, 104–115 (2015).

[21] J. Vehlow, Air pollution control systems in WtE units: An overview. *Waste Manag.* **37**, 58–74 (2015).

[22] L. Makarichi, W. Jutidamrongphan, and K.-A. Techato, The evolution of waste-to-energy incineration: A review. *Renew. Sustain. Energy Rev.* **91**, 812–821 (2018).

[23] S. Q. Aziz, H. A. Aziz, M. S. Yusoff, M. J. K. Bashir, and M. Umar, Leachate characterization in semi-aerobic and anaerobic sanitary landfills: A comparative study. *J. Environ. Manage.* **91**, 2608–2614 (2010).

[24] S. Mor, K. Ravindra, R. P. Dahiya, and A. Chandra, Leachate characterization and assessment of groundwater pollution near municipal solid waste landfill site. *Environ. Monit. Assess.* **118**, 435–456 (2006).

[25] B. P. Naveen, D. M. Mahapatra, T. G. Sitharam, P. V. Sivapullaiah, and T. V. Ramachandra, Physico-chemical and biological characterization of urban municipal landfill leachate. *Environ. Pollut.* **220**, 1–12 (2017).

[26] D. R. Caulton, P. B. Shepson, M. O. L. Cambaliza, D. McCabe, E. Baum, and B. H. Stirm, Methane destruction efficiency of natural gas flares associated with shale formation wells. *Environ. Sci. Technol.* **48**, 9548–9554 (2014).

[27] H. Zhao, *Methane Emissions from Landfills* (Columbia University, 2019). https://gwcouncil.org/m-s-thesis-methane-emissions-from-landfills/.

[28] Y.-T. Lu, Y.-M. Lee, and C.-Y. Hong, Inventory analysis and social life cycle assessment of greenhouse gas emissions from waste-to-energy incineration in Taiwan. *Sustain.* **9**(11), 1959 (2017).

[29] J. W. Sutherland, J. S. Richter, M. J. Hutchins, D. Dornfeld, R. Dzombak, J. Mangold *et al.*, Robin the role of manufacturing in affecting the social dimension of sustainability. *CIRP Ann.* **65**(2), 689–712 (2016).

[30] Z. Zhou, Y. Tang, Y. Chi, M. Ni, and A. Buekens, Waste-to-energy: A review of life cycle assessment and its extension methods. *Waste Manag. & Res.* **36**(1), 3–16 (2018).

[31] J. Song, L. Jin, C. Qian, and Y. Sun, *Economic, Social and Environmental Costs of the Waste-to-Energy Industry* (Oxford University Press, US, 2021).

[32] M. Haraguchi, A. Siddiqi, and V. Narayanamurti, Stochastic cost-benefit analysis of urban waste-to-energy systems. *J. Clean. Prod.* **224**, 751–765 (2019).

[33] M. L. Miranda and B. Hale, Waste not, want not: The private and social costs of waste-to-energy production. *Energ. Policy* **25**(6), 587–600 (1997).

[34] L. Messmann, V. Zender, A. Thorenz, and A. Tuma, How to quantify social impacts in strategic supply chain optimization: State of the art. *J. Clean. Prod.* **257**, 120459 (2020).

[35] C. Benoît-Norris and B. Mazijn (Eds.), *Guidelines for Social Life Cycle Assessment of Products* (United Nations Environment Programme (UNEP), 2009).

[36] C. Benoît-Norris, M. Traverso, S. Neugebauer, E. Ekener, T. Schaubroeck, G. Russo *et al.* (Eds.), *Guidelines for Social Life Cycle Assessment of Products and Organizations 2020* (United Nations Environment Programme (UNEP), 2020).

[37] C. Benoît-Norris, G. Vickery-Niederman, S. Valdivia, J. Franze, M. Traverso, A. Ciroth *et al.*, Introducing the UNEP/SETAC methodological sheets for subcategories of social LCA. *Int. J. Life Cycle Assess.* **16**, 682–690 (2011).

[38] L. V. Rasmussen, R. Bierbaum, J. A. Oldekop, and A. Agrawal, Bridging the practitioner-researcher divide: Indicators to track environmental, economic, and sociocultural sustainability of agricultural commodity production. *Glo. Env. Change* **42**, 33–46 (2017).

[39] M. J. Hutchins, J. S. Richter, M. L. Henry, and J. W. Sutherland, Development of indicators for the social dimension of sustainability in a U.S. business context. *J. Clean. Prod.* **212**, 687–697 (2019).

[40] Y. T. Chong, K. M. Teo, and L. C. Tang, A lifecycle-based sustainability indicator framework for waste-to-energy systems and a proposed metric of sustainability. *Ren. and Sust. Energy Rev.* **56**, 797–809 (2016).

[41] R. E. Miller and P. D. Blair, *Input-Output Analysis: Foundations and Extensions* (Cambridge University Press, Cambridge, United Kingdom, 2009).

[42] L. Mancini and S. Sala, Social impact assessment in the mining sector: Review and comparison of indicators frameworks. *Res. Policy* **57**, 98–111 (2018).

[43] G. Hardadi and M. Pizzol, Extending the multiregional input-output framework to labor-related impacts: A proof of concept. *J. Ind. Ecol.* **21**(6), 1536–1546 (2017).

[44] J. S. Richter, G. P. Mendis, L. Nies, and J. W. Sutherland, A method for economic input-output social impact analysis with application to U.S. advanced manufacturing. *J. Clean. Prod.* **212**, 302–312 (2019).

[45]   Waste Incineration and Public Health, *National Research Council (US) Committee on Health Effects of Waste Incineration: Social Issues and Community Interactions* (National Academies Press, Washington, DC, 2000).

[46]   M. Martuzzi, F. Mitis, and F. Forastiere, Inequalities, inequities, environmental justice in waste management and health. *Eur. J. Public Health* **20**, 21–26 (2010).

[47]   P. W. Tait, J. Brew, A. Che, A. Costanzo, A. Danyluk, M. Davis *et al.*, The health impacts of waste incineration: A systematic review. *Aust. N. Z. J. Public Health* **44**, 40–48 (2020).

[48]   X. Yuan, X. Fan, J. Liang, M. Liu, Y. Teng, Q. Ma *et al.*, Public perception towards waste-to-energy as a waste management strategy: A case from Shandong, China. *Int. J. Environ. Res. Public Health* **16**, 2997 (2019).

[49]   T. Cole-Hunter, F. H. Johnston, G. B. Marks, L. Morawska, G. G. Morgan, M. Overs *et al.*, The health impacts of waste-to-energy emissions: A systematic review of the literature. *Env. Res. Lett.* **15**(12), 123006 (2020).

[50]   J. L. Domingo, A. Bocio, M. Nadal, M. Schuhmacher, and J. M. Llobet, Monitoring dioxins and furans in the vicinity of an old municipal waste incinerator after pronounced reductions of the atmospheric emissions. *J. Environ. Monit.* **4**(3), 395–399 (2002).

[51]   J. L. Domingo, Human health risks of dioxins for populations living near modern municipal solid waste incinerators. *Rev. Environ. Health.* **17**(2), 135–147 (2002).

[52]   C. D. M. de Araújo, C. C. de Andrade, E. de Souza e Silva, and F. A. Dupas, Biodiesel production from used cooking oil: A review. *Renew. Sustain. Energ. Rev.* **27**, 445–452 (2013).

[53]   Y. Liu, C. Sun, B. Xia, C. Cui, and V. Coffey, Impact of community engagement on public acceptance towards waste-to-energy incineration projects: Empirical evidence from China. *Waste Manag.* **76**, 431–442 (2018).

[54]   J. S W. Levis, M. A. Barlaz, J. F. DeCarolis, and S. R. Ranjithan, A generalized multistage optimization modeling framework for life cycle assessment-based integrated solid waste management. *Environ. Model. Softw.* **50**, 51–65 (2013).

[55]   U. Sonesson, M. Dalemo, K. Mingarini, H. Jönsson, ORWARE — A simulation model for organic waste handling systems. Part 1: Model description. *Resour. Conserv. Recycl.* **21**, 17–37 (1997).

[56]   F. Ardolino, C. Lodato, T. F. Astrup, and U. Arena, Energy recovery from plastic and biomass waste by means of fluidized bed gasification: A life cycle inventory model. *Energy* **165**(Part B), 299–314 (2018).

[57]   P. Das, J.-C. P. Gabriel, C. Y. Tay, and J.-M. Lee, Value-added products from thermochemical treatments of contaminated e-waste plastics. *Chemosphere* **269**, 129409 (2021).

[58] J. Gong, B. Michalkiewicz, X. Chen, E. Mijowska, J. Liu, Z. Jiang *et al.*, Sustainable conversion of mixed plastics into porous carbon nanosheets with high performances in uptake of carbon dioxide and storage of hydrogen. *ACS Sustain. Chem. Eng.* **2**(12), 2837–2844 **(2014).**

[59] L. Matsakas, Q. Gao, S. Jansson, U. Rova, and P. Christakopoulos, Green conversion of municipal solid wastes into fuels and chemicals. *Elec. J. Biotech.* **26**, 69–83 (2017).

[60] A. Li, B. Antizar-Ladislao, and M. Khraisheh, Bioconversion of municipal solid waste to glucose for bio-ethanol production. *Bioprocess Biosyst. Eng.* **30**, 189–196 (2007).

Chapter 16

A Life Cycle Approach for Assessing the Impacts of Land-Use Systems on the Economy and Environment: Climate Change, Ecosystem Services, and Biodiversity

Miguel BRANDÃO

KTH — Royal Institute of Technology, Stockholm, Sweden

miguel.brandao@abe.kth.se

In view of the competing demands of land use to feed the growing population, sustain biodiversity, ecosystem services, and mitigate climate change, there is a clear need for a systematic approach for allocating land use with respect to economic and environmental objectives. This study formulates an integrated environmental and economic assessment of the global consequences of changing current land use in the UK with different land-use strategies for food, feed, fuel, timber, and carbon sink. Life cycle assessment (LCA) is used for the environmental assessment and a parallel economic assessment is integrated with LCA for the characterization of the main land-use strategies in the UK. The results indicate that changing land use and management on current cropland generally does not deliver improvement in all three criteria of mitigating climate and impacts on ecosystem service and biodiversity, while creating additional economic value. Expanding cropland onto set-aside and permanent grassland is more beneficial when crops are used for fuel or for carbon sink. Expansion onto set-aside grassland is largely undesirable if by arable cropping, but desirable by energy and forestry crops. The consequential assessment showed that indirect effects are relevant and ought to be considered when assessing land-use strategies.

1. Introduction

Concerns over global climate change, biodiversity loss, and degradation of ecosystem services have led to interest in using land in ways that mitigate these threats, particularly by growing various forms of energy crops to displace fossil fuels. In view of the competing demands on land to feed the growing population adequately, mitigate climate change, sustain biodiversity and ecosystem services, there is a clear need for a systematic approach for allocating land to the competing uses with respect to economic and environmental objectives. The purpose of this study is to formulate an integrated environmental and economic sustainability assessment for comparing different land-use strategies in the UK for food, feed, fuel, timber, and carbon sink.

While the first four functions of land are easily defined, the latter may require further explanation. With the purpose of mitigating climate change, carbon sequestration or biological carbon capture and storage (as opposed to geological, or fossil, carbon capture and storage) refers to a form of geo-engineering that consists in the biological removal (through photosynthesis) and storage of carbon from the atmosphere by trees. This land use decreases the atmospheric concentration of CO_2 while increasing the carbon stock in the biosphere and, thereby, acts as a carbon sink.

The comparison is carried out by analyzing the systems within a wide boundary, so that it includes consequences in other parts of the world. The focus is the cultivation of crops for different uses and the subsequent displacement they cause. This requires methodological developments for the estimation of substitution effects, i.e. the consequential changes happening as a result of each land-use strategy (e.g. product displacement and indirect land use changes).

Life cycle assessment (LCA) is used for environmental assessment. This requires the development of LCA methodologies and characterization factors for assessing land-use impacts on global climate change (due to temporary carbon storage), ecosystem services, and biodiversity. A parallel economic assessment is integrated with LCA, also using a life cycle perspective.

The application of the above methods allows for the achievement of the following objectives:

- To characterize, compare, and contrast the global environmental and economic consequences of alternative land-use strategies in the UK.
- To identify the land-use strategies in the UK that, on a global level, contribute the most to mitigating both climate change and impacts on the provision of ecosystem services and on biodiversity, while increasing global economic value.

2. Sustainability of Land Use: A Systems Approach

This section follows up on the concern identified in Section 1 regarding the prospects for increasing the supply of the various ecosystem services derived from land in limited availability.

By using a systems approach for the analysis of the environmental sustainability of land use and land-use change, Ref. [1] presents a background overview of the past and present links between land use, society, and environment (including in the UK).

Land use has changed dramatically throughout time in order to support the changing needs of an increasing human population. This has mainly consisted in the conversion of forests to cropland and grassland during the first agricultural revolution. The industrial revolution, coupled with the green revolution, allowed for unprecedented increases in yield, as advances in technology (e.g. synthetic fertilization, machinery and breeding) resulted in impressive increases in yield. However, these improvements have not eradicated hunger in the world and, despite some local and sporadic improvements, food insecurity (in both absolute and relative terms) is increasing.

Land cover and land-use changes are among the most significant impacts of human society on the environment, particularly on climate, ecosystems, and biodiversity.

Focus is given to the carbon budget, particularly the fluxes between the biosphere and the atmosphere (i.e. land-use change and land sink), which are disrupted by anthropogenic land use. The historical net effect on the carbon balance is substantial. The close link between the carbon and energy cycles highlights the impact on other species caused by disruptions to these cycles.

1. Provisioning Ecosystem Services
2. Regulating Ecosystem Services
3. Cultural Ecosystem Services
4. Supporting Ecosystem Services
5. Natural Capital
6. Human Capital
7. Social Capital
8. Manufactured Capital
9. Financial Capital
10. Human Health
11. Natural Resources
12. Natural Environment
13. Man-made Environment

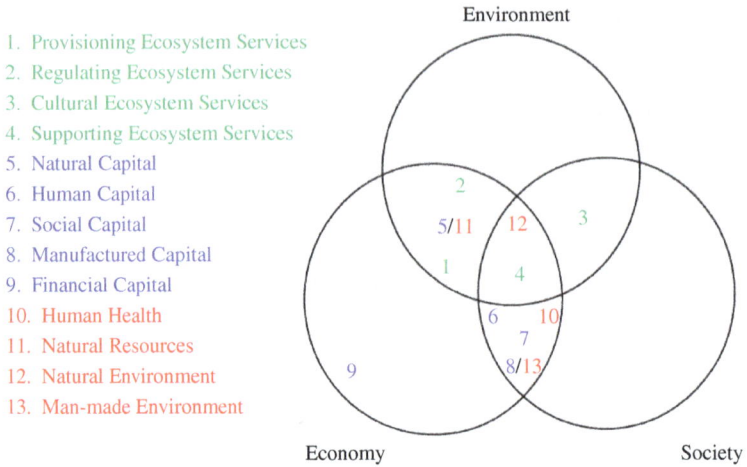

Figure 1.   Relationship between the different areas of protection commonly adopted in LCA and other popular frameworks in economics, ecology, and sustainability.

Ref. [2] identifies a multitude of system tools and frameworks, including the life cycle approach, for assessing sustainability. The multitude of conceptual frameworks and models that exist for assessing sustainability overlap and hence may possibly result in confusion and redundancy (see Figure 1). Ref. [1] integrates the concepts and terms used in these frameworks with those commonly used in LCA. As a result, a harmonized framework that bridges life cycle impact assessment (LCIA) with more widely accepted concepts and terms used in ecology and economics is proposed, allowing the convergence of different disciplinary frameworks.

In the proposed framework, a parallel life cycle economic assessment is merged with the environmental LCA, thereby enabling an integrated economic–environment assessment that is internally consistent.

3. Consequential Life Cycle Framework and Methodology for the Integrated Economic and Environmental Sustainability Impact Assessment of Land-Use Systems

The decision support analysis of land allocation to competing purposes is a complex endeavor due to the multitude of objectives aimed at.

Consequently, proper prioritization requires a systematic comparison of alternatives. This complexity is further enhanced by the multifunctionality of certain land uses.

Ways for using a life cycle approach to inform a systematic comparison of alternative land uses are explored in Ref. [3]. In a holistic framework, only a consequential life cycle approach is adequate, considering both upstream and downstream consequences, so that shifting of burdens between different components of the system (regardless of their geographical location) does not go unnoticed (see [4]).

A range of methodological issues relevant to assessing the consequential life cycle environmental and economic impacts of land-use strategies are identified in Ref. [3] and examined. In particular, Ref. [2] does a thorough analysis of the following four crucial methodological issues that, despite the recent methodological developments in this area, remain largely unresolved:

(1) *Estimation of land-use impacts on ecosystems*: These consist of the impacts from both land transformation and occupation on ecosystem services and biodiversity.
(2) *Estimation of land-use impacts on climate*: These consist of the impacts from both land transformation and occupation on the ecosystem's carbon stock.
(3) *Estimation of indirect land-use change (iLUC)*: The use of wide system boundaries that capture indirect land-use changes has been ignored in most LCA studies due to the popular adoption of attributional goals and scopes. This is a decisive consideration in the calculation of the carbon footprint of biofuels.
(4) *Decision support under multiple and conflicting objectives*: LCA has fallen short of supporting decision-making, due to the ambiguity given by a wide range of environmental considerations.

A review of the existing methods for assessing land-use impacts on ecosystems is conducted and their suitability and operability are assessed in [2]. Indicators associated with biomass, carbon balance, soil erosion, salinization, energy, soil microbial biota, and soil organic matter (SOM) are evaluated in Ref. [5]. The reason explaining the exclusion of the above issues from LCAs is partly due to the lack of robust LCIA methods and data quality.

The above methodological issues are addressed and a model with which land-use decisions may be assessed is proposed in Ref. [3].

The model relies on the development of the following operational approaches:

(1) Characterization factors for land-use impacts on ecosystem services and biodiversity
(2) Characterization factors for land-use impacts on climate
(3) New methods for the consequential assessment of land-use decisions
(4) The identification of the best strategies under a Pareto analysis framework.

The quantitative methodological developments that are proposed in Ref. [3] for the assessment of land-use impacts on climate change and on the provision of ecosystem services and biodiversity are based on the changes in carbon stocks of land-use systems. Here, impacts can be counted in LCA using the concepts of Ecosystem Carbon Stock (ECS) and its Human Appropriation (HAPECS). ECS represents the changes in the carbon balance in the area of the ecosystem under assessment due to human activity. It is expressed as tons C-year present or absent in the ecosystem (vegetation and soil) due to the land use studied and is selected as a promising standalone indicator because of its close association with most soil functions and biodiversity. Existing databases on terrestrial carbon stocks and land use enable this method to be applied robustly. As the HAPECS indicator was considered the most appropriate, characterization factors are developed in Ref. [2].

Different approaches for accounting for temporary carbon storage (or lack thereof) in land over a period of time are reviewed and the scope for their applications in LCA is explored in Ref. [6]. The results are very sensitive to the method chosen. The sensitivity of the results to the time period adopted is highly significant and therefore, a standard period, such as 100 years (GWP100) is recommended. In Ref. [3], the methods recommended are transferable to both GWP20 and GWP500; specifically the use of an updated equivalence factor based on the Moura-Costa method is proposed, and characterization factors are published in Ref. [2].

A consequential analysis is proposed in Ref. [3] for the estimation of iLUC, which is based on the substitution effects that arise from the outputs of the land-use strategies under study. As opposed to economic equilibrium models used for the same purpose, it takes a pure biophysical (rather than economic) displacement perspective.

In Ref. [3], the identification of the best strategies under a Pareto analysis framework is proposed, whereby only the strategies that represent an improvement from the baseline in all three criteria (climate, biodiversity and ecosystem services, and economic value) are considered further.

4. Integrated Assessment of Land-Use Strategies in the UK

The purpose of this section is to describe the application of the integrated environmental and economic framework, methods, and characterization factors that were developed in Refs. [2, 3] for the assessment and comparison of a wide range of possible land-use strategies in the UK — for food, feed, fuel, forestry, and carbon sink — not only in terms of their impacts on climate change, ecosystem services, and biodiversity but also in terms of the value they generate and for the identification of the best strategies in a Pareto optimal sense.

In Ref. [2], 224 scenarios under three sets of strategies were devised. The approach to consequential land-use change is applied to the scenarios by identifying the substitution effects from the different strategies. These include marginal suppliers, marginal land, marginal products, marginal feedstocks, changes in supply, and associated land area requirements. See Figure 2 for an example of one land-use strategy.

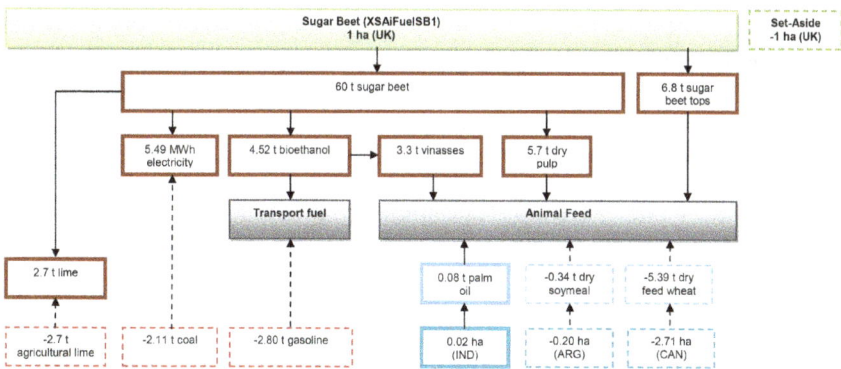

Figure 2. System delimitation of sugar beet expansion onto set-aside grassland for bioethanol and the associated displacements in Europe/USA and overseas (including iLUC).

4.1. *Strategies for current cropland*

4.1.1. *Land-use strategy A: Diversion*

Despite some relative improvements, diverting food or crop residues (straw or forestry residues) to feed or fuel does not deliver improvements in all three criteria simultaneously, with the exception of the case of wheat diversion from food to feed purposes. This is the only scenario in this strategy set that represents a Pareto improvement, whereby impacts on climate change and on ecosystem services and biodiversity decrease, while total economic value increases.

4.1.2. *Land-use strategy B: Extensification*

Extensification for either food, feed, or fuel always saves GHGs, with the exception of sugar beet for fuel. However, only organic wheat for food and feed and organic barley for feed satisfy all three criteria of mitigating GHG emissions and ecosystem services and biodiversity impacts while creating extra economic value.

4.1.3. *Land-use strategy C: Intensification*

Intensification for either food, feed, or fuel always saves GHGs, with the exception of oilseed rape for fuel. However, only intensive wheat and barley for feed satisfy all three criteria.

4.2. *Strategies for land expansion onto set-aside and permanent grassland*

Only intensive strategies present improvements in all three categories. Out of the arable crops, only intensive sugar beet production for ethanol results in Pareto improvements. Similarly, only intensive miscanthus and willow SRC for CHP proves beneficial in those three spheres. Forestry crops perform well, particularly Douglas fir whose different strategies for fuel or carbon sink are Pareto optima.

With only one exception, expansion of arable cropping is undesirable and, out of the perennial crops, miscanthus, willow SRC, ash, sycamore and silver birch, and Douglas fir are the only Pareto optimal strategies.

5. Discussion and Conclusion

This study shows that an integrated and holistic sustainability assessment is both possible and relevant and may be a suitable alternative to economic models and attributional LCAs. Out of the 228 scenarios, only 14 were identified as being sustainable cultivating wheat for food. Of those, only one strategy for biofuels was Pareto, that is, intensive sugar beet production on set-aside land. This points out that the most popular land-use strategy for climate change mitigation (i.e. biofuels) is not the most efficient, even if climate change is the sole concern. It also points out that, more often than not, changes in the background system negate any benefits that changes in the foreground system may have.

Instead, energy crops for electricity displace more GHGs than biofuels and contribute more to support ecosystem services and biodiversity. Not surprisingly, forestry crops for carbon sinks are better for both climate change and ecosystem services and biodiversity, but generate less economic value. Not all crops, regardless of the strategy employed, present opportunities for becoming more sustainable. The best option for improving sustainability on the existing arable land is limited to converting wheat for food to organic management and using it for feed.

Expansion onto set-aside land is where most Pareto optimal strategies lie. Here, intensive fuel production dominates, regardless of the crop used. Other than these strategies, only European larch for timber and Douglas fir for carbon sink pass all criteria for their attractive scores against ecosystem services and biodiversity and climate criteria, although the value generated by these two strategies is relatively insignificant.

At the extremes, carbon sinks by Douglas fir represent the best environmental use of land in terms of climate change mitigation and ecosystem services and biodiversity preservation while creating modest economic value, whereas ethanol from intensive sugar beet production generates the highest global net value while mitigating climate change a little.

Given the breadth of the consequential assessment, a large degree of uncertainty is involved (see Figure 3). The following modeling choices inevitably result in some form of uncertainty in the different phases of the LCA: functional unit, system boundaries, scenarios, consumption and production functions, equivalence between the different vegetable oils, choice of marginal producers, choice of marginal feedstocks, marginal yields, impact assessment methods, changes in ecosystem carbon stock,

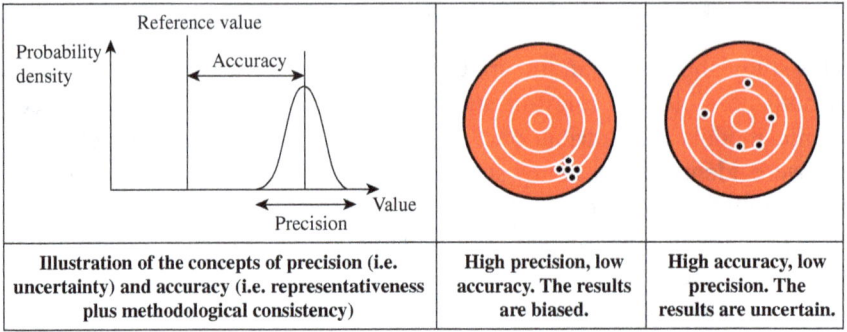

Figure 3.   Illustration of the concepts of precision and accuracy.

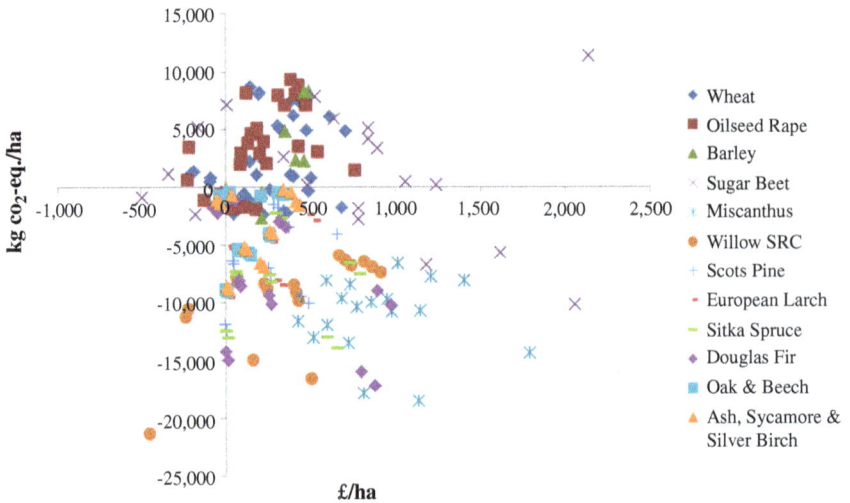

Figure 4.   Impacts on economic value added and climate change of all strategies.

choice of function and time in which land transformation impacts are allocated (see Figures 4–6 and Tables 1–2), time over which impacts are integrated, etc.

Given the multitude of objectives related to land-use policy (e.g. decreased climate and ecosystem services and biodiversity impacts, fuel security and redistribution of income to farmers), it is relevant that an integrated environmental and economic assessment is performed to identify the most sustainable land-use strategies. This has been done in an

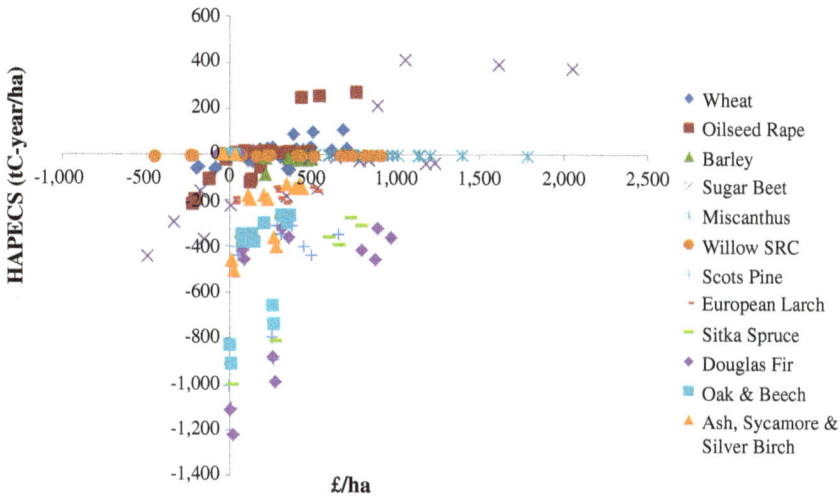

Figure 5. Impacts on economic value added and ecosystem services and biodiversity of all strategies.

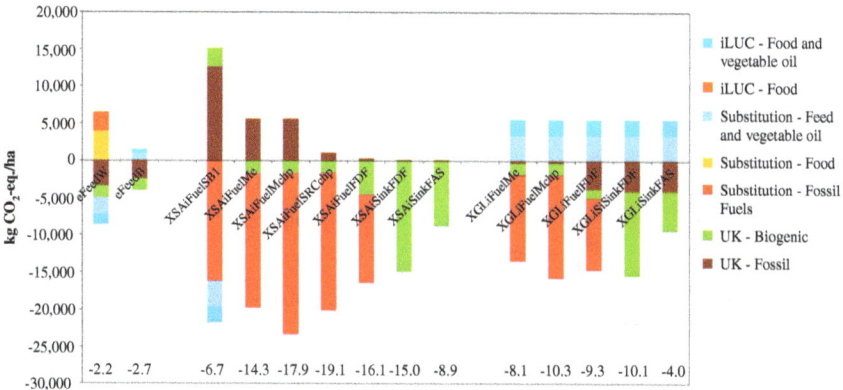

Figure 6. Climate impacts in the foreground and background systems of the Pareto optimal strategies.

approach where the different criteria (environmental and economic) are not weighted and traded as is common in neoclassical approaches, such as cost–benefit analysis. Nonetheless, an approach that quantifies, for example, externalities (such as carbon emissions) or ecosystem services, could follow this approach in an economic optimization model, despite the

Table 1.   Land transformation and occupation impacts.

	Transformation			Occupation		Total
Year	Annual Burden (%)	Cumulative Burden (%)	Annual Impact (tC-year)	Cumulative Impact (tC-year)	Impact (tC-year)	Impact (tC-year)
1	9.75	9.75	97.5	97.5	100	197.5
2	9.25	19.00	92.5	190.0	100	290.0
3	8.75	27.75	87.5	277.5	100	377.5
4	8.25	36.00	82.5	360.0	100	460.0
5	7.75	43.75	77.5	437.5	100	537.5
6	7.25	51.00	72.5	510.0	100	610.0
7	6.75	57.75	67.5	577.5	100	677.5
8	6.25	64.00	62.5	640.0	100	740.0
9	5.75	69.75	57.5	697.5	100	797.5
10	5.25	75.00	52.5	750.0	100	850.0
11	4.75	79.75	47.5	797.5	100	897.5
12	4.25	84.00	42.5	840.0	100	940.0
13	3.75	87.75	37.5	877.5	100	977.5
14	3.25	91.00	32.5	910.0	100	1010.0
15	2.75	93.75	27.5	937.5	100	1037.5
16	2.25	96.00	22.5	960.0	100	1060.0
17	1.75	97.75	17.5	977.5	100	1077.5
18	1.25	99.00	12.5	990.0	100	1090.0
19	0.75	99.75	7.5	997.5	100	1097.5
20	0.25	100.00	2.5	1000.0	100	1100.0

methodological constraints and the ethical limitations that this approach would represent.

It is exactly market failure in the allocation of agricultural resources and the resulting impact on sustainability that justifies regulation in agriculture and therefore policy support. This is the rationale for government intervention in agriculture, particularly its support to sustainable land systems or its taxation to farming systems that reduce sustainability (through, e.g. pollution). However, the current support for biofuel production is misguided by unclear objectives. While it may make sense in

Table 2. Land-use strategies for current cropland: ✓ represents Pareto optimal strategies, while ✗ represents other strategies that are not Pareto optimal.

Land Cover (Crop)	Diversion (A)			Extensification (B)			Intensification (C)			Total
	Food-to-Feed	Food-to-Fuel	Waste-to-Fuel	Food	Feed	Fuel	Food	Feed	Fuel	
Wheat	✗	✗	✗✗✗✗	✗	✓	✗	✗	✗	✗	1/12
Oilseed rape		✗	✗✗✗✗	✗		✗	✗		✗	0/9
Barley								✗		1/2
Sugar beet		✗		✗		✗	✗		✗	0/5
9 Forestry Crops			✗✗✗							0/18
			✗✗✗							
			✗✗✗							
			✗✗✗							
			✗✗✗							
			✗✗✗							
Total	0/1	0/3	0/22	0/3	2/2	0/3	0/3	0/2	0/3	2/46

meeting the objectives of security of both fuel supply and farmers' income, this research study indicates that for the mitigation of climate change and ecosystem services and biodiversity loss, biofuel strategies are highly inefficient and, therefore, largely redundant.

This research links very clearly with multi-objective linear programming and optimization and with multiple criteria decision analysis, at large. It would be very interesting to link this life cycle, bio-economic, economy-wide, and dynamic model with a multi-objective optimization model whereby all the relevant targets and variables are captured. An example of this could include the optimization of land use (and other agricultural resources) in the UK or Europe, with regard to environmental, economic, and social objectives. It could include several environmental constraints as well as targets that can be explored in combination or separately through scenarios. Additional research needs include, particularly, the validation and calibration of characterization models used in LCIA for land-use impacts on climate and on ecosystem services and biodiversity.

References

[1] M. Brandão, *Food, Feed, Fuel, Timber or Carbon Sink? Towards Sustainable Land Use: A Consequential Life Cycle Approach* (Springer, The Netherlands, 2021), p. 132.

[2] M. Brandão, *Assessing the Sustainability of Land Use, in Sustainability Assessment of Renewables-Based Products: Methods and Case Studies,* In: J. Dewulf, S. De Meester, and R. A. F. Alvarenga (Eds.), (John Wiley & Sons, Chichester, UK, 2015).

[3] M. Brandão, R. Clift, A. Cowie, and S. Greenhalgh, The use of LCA in the support of robust (climate) policy. *J. Ind. Ecol.* **18**(3), 461–463 (2014). https://onlinelibrary.wiley.com/doi/10.1111/jiec.12152.

[4] M. Brandão and L. M. i Canals, Global characterization factors to assess land use impacts on biotic production. *Int. J. Life Cycle Assess.* **18**(6), 1243–1252 (2012).

[5] M. Brandão, M. U. Kirschbaum, A. L. Cowie, and S. V. Hjuler, Quantifying the climate change effects of bioenergy systems: Comparison of 15 impact assessment methods. *GCB Bioenergy* **11**(5), 727–743 (2019).

[6] M. Brandão, *Integrated Environmental and Economic Assessment of Land-Use Strategies in the UK* (Lambert Academic Publishing, 2016), p. 84.

Index